T0134541

Springer Proceedings in Complexity

Springer Proceedings in Complexity publishes proceedings from scholarly meetings on all topics relating to the interdisciplinary studies of complex systems science. Springer welcomes book ideas from authors. The series is indexed in Scopus. Proposals must include the following: - name, place and date of the scientific meeting - a link to the committees (local organization, international advisors etc.) - scientific description of the meeting - list of invited/plenary speakers - an estimate of the planned proceedings book parameters (number of pages/ articles, requested number of bulk copies, submission deadline). Submit your proposals to: Christoph.Baumann@springer.com

More information about this series at http://www.springer.com/series/11637

Diane Payne • Johan A. Elkink • Nial Friel
Thomas U. Grund • Tamara Hochstrasser
Pablo Lucas • Adrian Ottewill
Editors

Social Simulation for a Digital Society

Applications and Innovations in Computational Social Science

 Springer

Editors
Diane Payne
School of Sociology
Geary Institute Dynamics Lab
University College Dublin
Dublin, Ireland

Nial Friel
School of Mathematics and Statistics
University College Dublin
Dublin, Ireland

Tamara Hochstrasser
School of Biology
and Environmental Science
University College Dublin
Dublin, Ireland

Adrian Ottewill
School of Mathematics and Statistics
University College Dublin
Dublin, Ireland

Johan A. Elkink
School of Politics
and International Relations
University College Dublin
Dublin, Ireland

Thomas U. Grund
School of Sociology
University College Dublin
Dublin, Ireland

Pablo Lucas
School of Sociology
Geary Institute Dynamics Lab
University College Dublin
Dublin, Ireland

ISSN 2213-8684   ISSN 2213-8692   (electronic)
Springer Proceedings in Complexity
ISBN 978-3-030-30300-6   ISBN 978-3-030-30298-6   (eBook)
https://doi.org/10.1007/978-3-030-30298-6

© Springer Nature Switzerland AG 2019
This work is subject to copyright. All rights are reserved by the Publisher, whether the whole or part of the material is concerned, specifically the rights of translation, reprinting, reuse of illustrations, recitation, broadcasting, reproduction on microfilms or in any other physical way, and transmission or information storage and retrieval, electronic adaptation, computer software, or by similar or dissimilar methodology now known or hereafter developed.
The use of general descriptive names, registered names, trademarks, service marks, etc. in this publication does not imply, even in the absence of a specific statement, that such names are exempt from the relevant protective laws and regulations and therefore free for general use.
The publisher, the authors, and the editors are safe to assume that the advice and information in this book are believed to be true and accurate at the date of publication. Neither the publisher nor the authors or the editors give a warranty, express or implied, with respect to the material contained herein or for any errors or omissions that may have been made. The publisher remains neutral with regard to jurisdictional claims in published maps and institutional affiliations.

This Springer imprint is published by the registered company Springer Nature Switzerland AG
The registered company address is: Gewerbestrasse 11, 6330 Cham, Switzerland

This book is dedicated to Assoc. Prof. Diane Payne, our dear colleague who passed away in early 2019. The Social Simulation Conference 2017 in Dublin, which she was instrumental in organizing, was subtitled: Social Simulation for a Digital Society. This motto expressed Diane's conviction that sociology would gain from using computer simulation and taking advantage of the increasing availability of data. Throughout her life, particularly during her tenure at University College Dublin, she promoted the use of quantitative methods in sociology as the Director of the UCD Dynamics Lab at the UCD Geary Institute for Public Policy and as Head of the School of Sociology. She was passionate about the opportunities offered that allowed and facilitated new collaborative research approaches between scientists, industry, and government – while involving different disciplines within UCD and other universities worldwide. She kept an open mind and an insatiable curiosity that allowed to constantly further develop analytical and modeling approaches. Through the European Social Simulation Association, she developed a professional network that shared her ethos, and she made sure that this research network did benefit from every opportunity available to staff and students. She was a pioneer in her field, and she is dearly missed. We are sure her professional legacy will continue to inspire the work of scientists across the world. Thank you, Diane.

Contents

Chapter 1
Social Simulation for a Digital Society: Introduction

Diane Payne, Johan A. Elkink, and Thomas U. Grund

An increasingly popular strand of social science research attempts to understand social facts (e.g. segregation, social inequality, cooperation, opinions, social movements) not merely by relating them to other social facts, but rather by detailing how relatively simple interactions between individuals and groups (agents) combine and lead to the emergence and diffusion of social patterns (Squazzoni 2012; Macy and Willer 2002). Simulation models to investigate such interactions have been around for decades in the social sciences, for example Schelling's (1978) model of social segregation or Axelrod's (1986, 1997) model of the evolution of cooperation, but only became more common as the computing power accessible to the typical social scientist increased.

A computational agent-based model is a model in which the patterns are studied that result from the interaction between large numbers of actors on the basis of a relatively simple set of behavioural assumptions, simulated in a computer environment. For example, some simple assumptions about the likelihood that a fish of a certain size will eat another fish, the likelihood that a fish will reproduce, and the likelihood that a fish will naturally die, can be modelled in a computer simulation, the analysis of which can provide useful insights in the ecology of fish, which are difficult to trace using other methods of modelling or simulation (DeAngelis and Rose 1992). These models are generally based on assumptions of non-linear relations between variables, due to the fact that actors both create or form their environment, while their environment affects their individual behaviour. The behaviour of

D. Payne
School of Sociology, University College Dublin, Dublin, Ireland

Geary Institute Dynamics Lab, University College Dublin, Dublin, Ireland

J. A. Elkink (✉)
School of Politics and International Relations, University College Dublin, Dublin, Ireland
e-mail: jos.elkink@ucd.ie

T. U. Grund
School of Sociology, University College Dublin, Dublin, Ireland
e-mail: thomas.grund@ucd.ie

© Springer Nature Switzerland AG 2019
D. Payne et al. (eds.), *Social Simulation for a Digital Society*, Springer
Proceedings in Complexity, https://doi.org/10.1007/978-3-030-30298-6_1

an individual agent is thus dependent on that of many other agents previously, which creates complex patterns not easily deductible from the individual rules of behaviour of the agents.

Such an approach overcomes the often falsely assumed dictum that individuals (or observations) are independent from each other. What is often regarded as a nuisance in many statistical analyses receives a full spotlight in social simulation studies. Social influence, feedback loops, tipping points, unintended consequences and the emergence of social phenomena from the 'bottom-up' take pivotal roles. Individual-level behaviours are investigated in the light of preceding social conditions (macro-micro relations).

At the same time, emphasis is put on how the behaviours of individuals combine and generate the social outcomes we observe (micro-macro relations). The effect of fairly straightforward interactions between individual members of a mass of people can have very complicated and often unexpected effects on the mass behaviour as a whole. For example the simple behaviour of a car driver, slowing down for cars in front of him or her and speeding up when there is a chance can, given different initial speeds of different cars on the road, easily lead to traffic jams. Simply slowing down for people in front does not trivially lead to traffic jams, yet such behaviour, given the diversity in speeds, does have this effect. Moreover, placing traffic lights on a road even without any crossroads can reduce the chances of a traffic jam because they have the effect of homogenizing the speeds of the cars – they all start at the same time at similar distances from each other when the light turns green. Therefore the intuitively contradicting idea of stopping cars to avoid traffic jams can actually be quite successful. This example illustrates in a simplistic way how individual behaviour can have unexpected macro effects and this link between local, individual behaviour and global, macro behavioural patterns thus deserves attention in social science research.

Computer simulations rewind history, investigate how social processes unfold and how starting conditions or interaction structures influence social outcomes. This breaks with traditional conception and thinking, emphasizes social dynamics and disconnects the sizes of causes and consequences.

Social simulation is a powerful tool to understand the macro-implications of micro-level dynamics. But it also allows the simulation of what-if scenarios. How would the world look like if a certain policy would be implemented? How would a market change if interaction rules for market participants would be altered? How would traffic change if a train line would be built? Answering such questions are not trivial in highly dynamic and complex social systems. Small changes can have huge implications and the sizes of causes and effects are often not proportional to each other anymore. Furthermore, real-world experiments are not always feasible or too costly. One cannot simply try out different locations of a subway stop in the real word, but one can simulate different scenarios.

As more and more scholars embrace computer simulations in the social sciences, there is also an expanding academic community in this field. This volume builds on a variety of contributions, first presented at the European Social Simulation Association Conference, hosted in Dublin in 2017. These contributions vary from philosophical considerations underlying this methodology, to methodological and technical contributions, to applications in particular in the domain of spatial social dynamics.

Chapters 2–4 deal with philosophical and technical considerations in social simulation studies. In the first contribution, Lia ní Aodha (Chap. 2) discusses the use of qualitative data to inform agent-based models. Her critical reflection provides a discussion of the ontological assumptions in agent-based modelling and qualitative research in order to outline potential incompatibilities. In Chap. 3, Jonathan Thaler and Peer-Olaf Siebers develop a classification of agent update-strategies – i.e. the timing and sequencing of updates and messaging between agents. Applying different update-strategies to well-known models, they illustrate the impact a modelling decision for one or another update-strategy might have. And then, Patrick Taillandier, Mathieu Bourgais, Alexis Drogoul and Laurent Vercouter (Chap. 4) provide a contribution where an existing modelling framework is altered to allow for parallelization. Using examples from the platform GAMA, they demonstrate how parallelization can be achieved and how much faster it makes simulations.

Chapters 5–8 apply social simulations to norm diffusion and collective action. Kyle Bahr and Masami Nakagawa (Chap. 5) develop a model where network ties are created through communication and continued communication strengthens network ties. Communication is affected by agents' level of influence, which is set by the centrality in the communication network. In Chap. 6, Emiliano Alvarez and Juan Gabriel Brida model individual agents on a regular grid, who change their opinion, based on personal preferences, neighbour's opinions, and random mutation. Christopher K. Frantz and Amineh Ghorbani (Chap. 7) models the impact of inequality on the sustainability of protest. While it is – that inequality matters for the initiation of protest, much less is known about its impact on sustainability. In Chap. 8, Oswaldo Terán, Christophe Sibertin-Blanc, Ravi Rojas and Liccia Romero expand on an existing model to investigate the interests and inter-play between different actors in the potato seeds market in Venezuela. Their simulations give insights into different policies and configuration options that would create a fairer market.

Chapters 9–14 provide application of social simulations to geography and urban development. Johannes Weyer, Fabian Adelt and Sebastian Hoffmann (Chap. 9) model traffic behaviour on a network of connections with nodes such as homes and work locations, and different parameters for agents that affect route selection. Demonstrating the usefulness of their model, while allowing for the addition of different types of actors, it shows the impact of different governance regimes. In Chap. 10, Liu Yang, Koen H. van Dam, Bani Anvari and Audrey de Nazelle provide an agent-based model to simulate traffic in Beijing, which is then compared to actual traffic data from Google Maps. In particular, they focus on the upgrade of a train line that cross-cuts a large part of Beijing. In Chap. 11, Hideyuki Nagai and Setsuya Kurahashi present an agent-based model for arranging transport methods and leisure facilities in different ways in a commuter town and work centre. Their model identifies measures that could affect compactification, car use and CO_2 emissions. Laura O. Petrov, Brendan Williams and Harutyun Shahumyan (Chap. 12) provide a discussion of engagement between stakeholders, policy-makers, and modellers in the context of urban and rural development modelling in a particular region in Ireland. Gillian Golden (Chap. 13) also focuses on Ireland and discusses procedures that can be used to create a synthetic population of children in Irish schools that match marginal statistical data available at regional or school level, using simulated annealing and various proposed goodness of fit measures. And lastly, Martina Neuländtner, Manfred Paier and Astrid Unger (Chap. 14) explore the impact of increased collaboration between firms as assessed by patent quantity. Increasing collaboration does not have a positive impact per se, since it is difficult to find suitable partners, but increasing national-international collaboration increases opportunities for diversity and avoids lock-in, which leads to innovation and more patents.

Sadly, one of the organizers of the European Social Simulation Conference 2017 and the main editor for this volume – Diane Payne – unexpectedly died in early 2019. Diane was a prominent figure in social simulation studies, both in Ireland, but also abroad. We wanted to finish publishing this volume in her name and dedicate it to her. She will be missed.

References

Axelrod, R. (1986). An evolutionary approach to norms. *American Political Science Review, 80*(4), 1095–1111.

Axelrod, R. (1997). *The complexity of cooperation: Agent-based models of competition and collaboration*. Princeton: Princeton University Press.

DeAngelis, D. L., & Rose, K. A. (1992). Which individual-based approach is most appropriate for a given problem? In D. L. DeAngelis & L. J. Gross (Eds.), *Individual-based models and approaches in ecology. Populations, communities and ecosystems* (pp. 67–87). New York: Chapman & Hall.

Macy, M., & Willer, R. (2002). From factors to actors: Computational sociology and agent-based modeling. *Annual Review of Sociology, 28*, 143–166.

Schelling, T. (1978). *Micromotives and macrobehavior*. New York: W.W. Norton.

Squazzoni, F. (2012). *Agent-based computational sociology*. Chichester: Wiley.

Part I
Philosophical and Technical Considerations

Chapter 2
Ontological Politics in a World of Political Ontologies: More Realistic (Human) Agents for the *Anthropocene*?

Lia ní Aodha

"VARIETY DISAPPEARS when subjected to scholarly analysis."

(Feyerabend 2001, p. 12)

Introduction

Given the "character of calculability" of the twentieth century (Mitchell 2002, p. 80), and the widespread propensity to *hyper*-quantification (Denzin 2017), alongside the mounting evidence with respect to the societal and ecological damages this has entailed (Pilkey and Pilkey-Jarvis 2007), adopting a qualitative approach to building agent-based models (ABMs) seems downright reasonable. In this sense, "getting away from numbers" (Yang and Gilbert 2008, p. 275) would certainly appear to have some merit. Nevertheless, this effort does invite critique. Some of these relate to the practice of modelling itself, whilst others are of a more general nature and are similar to those that have arisen in other collaborative research endeavours. In this sense, they may be placed within the context of a broader politics of representation, and of knowledge. It is argued here that consideration of these overlapping issues necessitates, in the first instance, reflection on the wider contextual backdrop against which these politics are unfolding, and within which complex computer simulations are gaining increasing precedence. Subsequently, whilst drawing attention to the embeddedness of *all* knowledge, including that which is operationalised through complex simulations, a number of philosophical and sociological critiques of agent-based modelling (ABM) are made. Here, questions are posed with respect to the assumptions underpinning ABMs, and the level of simplification they entail. In turn, the case is made for the importance of situated knowledge, whilst the *potential* colonising effect of complex computer simulations is both highlighted and cautioned against. Notwithstanding this, an overarching call for pluralism is put forward.

L. ní Aodha (✉)
Centre for Policy Modelling, Manchester Metropolitan University, Manchester, UK

© Springer Nature Switzerland AG 2019
D. Payne et al. (eds.), *Social Simulation for a Digital Society*, Springer Proceedings in Complexity, https://doi.org/10.1007/978-3-030-30298-6_2

Complex Problems and the Integration Imperative

In the spirit of this argument, context matters. Backgrounding this critique is an academic and policy environment that, very much focused on complex (social, eco-logical, socio-ecological) problems, increasingly calls for broad collaboration, drawing across (and beyond) disciplines – from the natural and social sciences to the humanities, and further. These calls themselves, are situated within a wider social, political and ecological *reality* that certainly requires collaboration and, arguably, further extension of that collaboration to different traditions of thought entirely. Given this, this call could (should?) be read as one that requires a plurality of views, and knowledge. The very real anxieties (Robbins and Moore 2013) of this post political moment[1] — increasingly known, though not without contention, as the Anthropocene — require pluralist, non-reductive strategies (Blaser 2009, 2013; Castree 2015; de Castro 2015; Klenk and Meehan 2015; Lövbrand et al. 2015).[2] In this sense, they might logically be read as necessitating methodological, alongside disciplinary, plurality.

Despite this, these calls and the manner in which they are being answered (once one scratches the surface) seem to betray a polemic, flabby, and defensive character (Bernstein 1989).[3] Though couched in collaborative dressing, they have been charged with displaying a methodological monistic hue, an underlying integration imperative (Castree et al. 2014; Klenk and Meehan 2015); with the broad focus in many spaces remaining narrowly centred on the individual (Castree 2015). Far from a fundamental rethink of the *enlightened* — predominantly quantitative, predictive, instrumental — approach to designing nature (Pilkey and Pilkey-Jarvis 2007, pp. 192–193), the guise of social science and humanities work that has gained most salience has been that which has adopted "…quantitative, behavioural approaches that fit the bill of a supposedly 'objective' representation…" (Pellizzoni 2015, p. 11). Whilst the picture sketched here is perhaps unsurprising, in the context of a re-emerging and narrowly defined call for "scientifically based research" and

---

[1] Post-politics is taken here to denote the largely consensual vision or arrangement, within which the current institutional setup, problem framings, and proposed solutions are taken as a given (for a more explicated reading see, for example, Ranciere 2001; Swyngedouw and Ernstson 2018; Žižek 2017).

[2] The contentiously proposed term for the new geological epoch that the Earth has allegedly entered, as a consequence of human (singular and homogenous) activity (Bavington 2011; Lorimer 2017). Interpretations and mobilisations of the concept are many, and an array of alternative (more accurate) renderings exist (for example, Lorimer 2017; Moore 2016; Swyngedouw and Ernstson 2018).

[3] Bernstein (1989, p. 15) laid out a number of different types of pluralisms: *Fragmentary pluralism* is one in which we fall back into our silos, only willing to communicate within our own safe 'thought collectives' (even if they are undisciplined). *Flabby pluralism* is where our pluralisms amount to "little more than superficial poaching". *Polemical pluralism* is where appeals to plural-ism "become little more than an ideological weapon to advance one's own orientation", without any real readiness to take others seriously. *Defensive pluralism*, again displays little real willing-ness to engage beyond mere lip service.

"evidence-based policy" (Maxwell 2004, p. 35; Saltelli and Giampietro 2017), whereby the emphasis is trained upon a narrow rendering of something called "the human dimensions" (Bavington 2011, p. 18; Castree 2015; Castree et al. 2014; Pellizzoni 2015), such a hue and focus sits uneasy in a world of ontological politics (Pellizzoni 2015). Ontology matters (Epstein 2016a). And all ontologies are political![4]

Agent-Based Models and Qualitative Inquiry

"Qualitative evidence has often been seen as "unscientific", critiqued as: subjective, biased, unreliable and context-specific. These critiques are not without merit – qualitative evidence *does* have its difficulties – but it also has its own advantages and the difficulties are not sufficient to justify ignoring it" (Edmonds 2015, p. 1).

How does this relate to the question as to whether or not there are critiques to be levelled at the ongoing attempt(s) to utilise qualitative data to inform ABMs? As indicated, there *does* seem to be good reason to suggest that we need to start thinking more qualitatively (Pilkey and Pilkey-Jarvis 2007, pp. 192–193), and ABM does appear to be a good vehicle for incorporating qualitative evidence into complex models. ABM, it is argued, is particularly suited to encoding aspects of qualitative data, thereby allowing for the avoidance of unnecessary quantification (Edmonds 2015). At the same time, qualitative data may go a long way to capturing the micro-level data that is required in building *a* model (Edmonds 2015), and can be hugely beneficial in, subsequently, validating *the* model, given its ability to capture also some contextual macro features which may be verified with the outcomes generated by *that* model (Yang and Gilbert 2008).

So in this sense, yes – *narrative textual evidence* can provide a good basis for informing the behavioural rules of virtual agents, and more. The work in this area verifies this, and most certainly, such an approach can lend a certain realism to a model (e.g. Bharwani et al. 2015). Given that, to date, evidence-driven models remain, somewhat, thin on the ground, adopting a qualitative strategy may go some way to closing this gap (ibid), and satisfy the need to embed models in empirical data (Boero and Squazzoni 2005). Further, and with respect to collaboration, given its capacity to bridge the quali-quantitative divide (Squazzoni et al. 2014), ABM, it is suggested, may be a promising instrument through which traditional disciplinary boundaries might be traversed, thereby potentially offering dividends in terms of plural research endeavours (Squazzoni 2010).

---

[4] Ontology is used here to denote the assumptions through which we apprehend and depict "reality" (Kohn 2015), with the underlying premise being that when paired with the word politics signifies that the "real" is not necessarily a given, but rather is historically, culturally, and materially situated (Mol 1999).

Notwithstanding this, arguably, a lot of the above argument echoes some of the *potential* dangers of such efforts. For example, taking Edmonds' (2015) quote above, it is true that for some academics, qualitative and quantitative evidence are incommensurable – that's ok, this is in *some ways* simply reflective of different ways of understanding and seeing the world.[5] Different ways of making sense of the world are important, and it is not at all clear that commensurability is required. Thus, while ABM may be a possible vehicle for bridging this gap, whether or not finding a way where both approaches can be folded in together is or should be the objective for either side of the divide is certainly open to question. Put mildly, in the *Anthropocene*, the politics of evidence and representation matters a great deal, which merits reflection when thinking about questions of commensurability and incommensurability, and with respect to the kind of world that is rendered legible through ABM. These factors are not inconsequential when discussing the potentialities of plural research endeavours. Further, their consideration seems especially pertinent given that there is evidence to suggest that ABM is gaining traction not only across the social sciences (Castree et al. 2014; Squazzoni 2010), but beyond (Government Office for Science 2018; Pellizzoni 2015).

A Philosophical and Sociological Critique

Though there does seem to be a decent amount of evidence that there are gains to be had — certainly in terms of more realistic models — from using qualitative data to inform ABMs, a number of philosophical and sociological critiques may be levelled at ABM, which could render it an untenable approach for many perspectives working within (and around) the social sciences and humanities. In this respect, two issues in particular are raised here: ABM's intentional or unintentional, and often mentioned, but seemingly persistent propensity towards methodological individualism, and its demand for simplification.[6] The first point here raises questions with respect to the ontological base of models, whilst the second clearly entails limitations in terms of the contextual, deeply situated analysis that is at the core of much qualitative research. Together they elicit a number of questions with respect to the kind of world that is being depicted by ABM, and open discussion for the kinds of depictions ABM might provide otherwise (Holbraad et al. 2014).

---

[5] Having said that, part of the "ontological turn" across the social sciences has entailed a shift from questioning different viewpoints to posing questions with respect to the reality of different worlds i.e. with respect to ontological multiplicity, rather than epistemic multiplicity (Pellizzoni 2015).

[6] Methodological Individualism – the view that human individuals are the sole, unique, and ultimate constituents of social reality to which all else is reducible (Hay 2013).

ABM's Methodological Individualism

Ontological decisions regarding the kinds of entities we assume to exist, or whether we decide to carry out our inquiry in terms of identities, individuals, collectives, states, regimes, systems, or something else, reflect certain assumptions about 'reality' (Hay 2013). These choices, even if they are only implicit (or we have not even thought about them), have epistemological, methodological, and practical consequences (Hay 2013). In this respect, it has been suggested that ABM and qualitative research, ontologically and epistemologically speaking, are not very removed from one another; with context, time, mechanisms, processes, and sequences of events being important to both (Yang and Gilbert 2008).[7] Though this may be the case, a number of ontological misconceptions remain widespread in modelling, concerning the relation between macro and micro properties (Epstein 2013, 2016b). Here, the long-standing argument made is that many models display a level of methodological individualism (MI), and a leaning towards a unidirectional micro-to-macro level emergence (Conte et al. 2001; Epstein 2013; O'Sullivan and Haklay 2000; Venturini et al. 2015) that may certainly make some social scientists uncomfortable (e.g. see Bourdieu 1989; Emirbayer 1997; Knorr-Cetina 1988).

Indeed, there *is* much sociological evidence to suggest that such a stance is (at best) limited (Conte et al. 2001; Sawyer 2005), even for basic 'facts' about groups of people (Epstein 2016b). For example, ongoing work within the subfield of social ontology (as elsewhere) makes the case that group intention and action can, and often does depend on more than the individuals within the group – external forces, power, hierarchies, meso- and macro-forces etc. matter (ibid). From an array of perspectives, non-human materialities matter (Elder-Vass 2017; Morton 2013, 2017). Indeed, similar arguments, with respect to *new*-materialisms, have long held traction in and around science and technology studies (STS), and political ecology (e.g. Callon 1984).[8] Essentially, the premise of MI, whether ontological or explanatory (as is sometimes posited), is questionable (Epstein 2016b). Related premises of weak individualism or structural individualism (Hedström and Ylikoski 2010; Marchionni and Ylikoski 2013) are equally dubious.[9]

Despite this, however, many ABMs seem stuck in and around this assumption. What is less clear entirely, however, is whether this MI (taken as sound or otherwise), is grounded in an ontological commitment.

---

[7] This statement merits clarification. In terms of underlying assumptions qualitative research is methodologically diverse. Thus, one may reasonably highlight here that some traditions of qualitative research might be quite removed — ontologically and epistemologically speaking — from the representational economy of agent-based models.

[8] From such a vantage point, matter itself is rendered agential. Further, the bounded organism is not the unit of study, rather the focus is on assemblage (MacLure 2017).

[9] Recent discussions of this kind have come from AS (Bulle and Phan 2017; Hedström and Ylikoski 2010). However, Little (2012) has argued that AS, despite postulations with respect to structural individualism, seems to be explicitly grounded in MI, with some possible commitment to methodological localism.

Although *all* research is underpinned by epistemological and ontological assumptions, a lot of models are not explicit about the worldviews and assumptions underpinning them. As Epstein (2013) has highlighted, even though it may not be glaringly obvious that ontological assumptions are important in ABM, they are certainly there – even if only implicitly. In this respect, there seems to be an affinity between the explicit methodological assumptions underpinning the mechanistic perspective of analytical sociology (AS) (Bulle and Phan 2017; Hedström and Ylikoski 2010; Little 2012) and those apparent in ABM.[10] Indeed, Hedström and Ylikoski (2010) have professed such a kinship. However, whether this affinity is bidirectional seems open to question. Put another way, it is unclear whether the MI displayed by many ABMs is simply a symptom of an approach that has "grown up" with a focus on generative mechanisms (Marchionni and Ylikoski 2013) and an ad hoc style, or whether it reflects a broad-based acceptance of the same premises as AS (e.g. Boero and Squazzoni 2005).

A number of different suggestions have been put forth with respect to these issues. Sawyer (in Conte et al. 2001), for example, has suggested that ABM displays this MI largely due to unquestioned assumptions, rather than pragmatic considerations, or empirical evidence. The author further suggests that this is possibly a feature of an earlier relation with economic theory (one does not have to look far to find ABMs with simplistic economic agents, or their bounded cousins), with the field of artificial intelligence (AI), and of cultural biases towards individualistic thinking, more generally. In short, this appears to be an unquestioned assumption rather than a foundational argument (Sawyer, in Conte et al. 2001), as it is with AS. Similarly, Epstein (2013) has argued that (computationally) ABM is not inherently individualistic (i.e. this is not a limitation that is built into ABM, per se). As such, this is something that can perhaps be overcome. In relation to this, however, Epstein (2013) has highlighted that given some of the issues surrounding this are in many instances overlooked completely by the researcher, models often don't succeed in avoiding even the crudest forms of individualism.

This discussion raises a number of issues relating to the embeddedness of *all* knowledge, and the consequentiality of failing to reflect on our own biases, or our assumptions more generally (even if we do not consider them to be situated). The points raised by Sawyer (in Conte et al. 2001), for instance, give countenance to the argument that context is an important determinant of the manner in which social data (or indeed any data) is rendered (Hacking 1990). How we apprehend and depict the world is grounded "within a larger context of what the individual is, and of what society is" (ibid, p. 4). Here, one might make the case that our predisposition to think at the individual level might reasonably be considered an *a*ffect of our own embeddedness within the current hegemonic socio-natural configuration.[11] With

---

[10] The ontological base of AS is open to many of the same critiques made here (e.g. see Little 2012). That "it makes quite a difference whether the world is viewed as a machine or as a turbulent stream" (Kwa 1994, p. 387) is worth considering here also.

[11] Margaret Thatcher's famous quip: "There is no such thing as society. There are individual men and women…" comes to mind here.

respect to both this and the points raised by Epstein (2013), suffice to say that if you fail to recognise that your knowledge is situated, you are unlikely to reflect on it (Berg 2001; Haraway 1988; Rose 1997).

Whichever the case, the proposition that individuals are "a stable and unproblematic source of social action" or "causal agents who produce, mediated by their dispositions and beliefs, a steady flow of social phenomena" (Knorr-Cetina 1988, p. 24) is at odds with *a lot* of perspectives across the social sciences and humanities. Likewise, *any* kind of micro-macro dualism or determinism is going to be problematic, for those sharing the view that the macro and micro cannot be ontologically separated at all, but rather are co-constituted, and this is the case even in an actor-oriented approach that prioritises individual meaning and action (Long and Long 1992). Further, the aforementioned question of embeddedness certainly arises (Granovetter 1985; Polanyi 1944), whilst a relational theorist will reject the notion outright that one can posit discrete, pre-given units such as the individual or society as the definitive starting point of social analysis (Bourdieu 1989; Emirbayer 1997). Indeed, Venturini et al. (2015, p. 3) go so far as to make the claim that "the *last* thing" social scientists need "are models that break them in micro/macro oppositions", highlighting that empirical evidence shows that social structures do not simply jump up from micro interactions, but rather there is a dialectical relationship of constant flux between and among both of these levels.

Simplification Versus Messy Reality

A further issue to be raised with respect to these models, and which is done so with the recognition that efforts to use qualitative evidence to inform ABMs represent an attempt to closing this gap, is the level of simplification that is required in ABM. In this respect, the points which have been raised in relation to the ontological base of ABM suggest a degree of (rectifiable) oversimplification. However, the level of simplification demanded of formal modelling more generally, is a line of inquiry worth following in considering how far qualitatively driven models might go, with respect to rethinking how we conceptualise and design nature.

In terms of representation, all formal modelling endeavours require a level of abstraction that is quite distinct from, for example, the narrative renderings of the world that qualitatively driven models seek to draw on. That said, it is recognised that ABM fares better than other models on this charge. For instance, unbounded from the constraints of analytical mathematics in a manner that is distinct from other formal models, ABM *does* have the capacity to allow for a "messier" representation (Squazzoni 2010; Squazzoni et al. 2014). Given its ability to deal with qualitative data, heterogeneity, and environmental features (ibid) ABM makes possible a degree of "ontological correspondence" with the real world that other models struggle with (Squazzoni 2010, p. 199). One might highlight here, however, that regardless of what the features of ABM make possible, the discussed ontological misconceptions that are prevalent in ABM call into question the degree of current correspondence. In terms of simplification versus messy realities, one might also highlight that ABM still suffers a number of limitations vis-à-vis the contextual, deeply situated analysis that is central to much qualitative research, which does call into question the manner in which these two ways of grappling with the world might best be "coupled".

O'Sullivan and Haklay (2000), for instance, highlight that within ABM there is a strong commitment to minimal behavioural complexity, in order to make the process of modelling feasible, and the resulting model understandable. Similarly, Venturini et al. (2015, pp. 1–2) highlight that models often entail a great degree of simplifying agents, their interactions and emergent structures, with the objective being to *fit* them, raising the issue that from "a methodological viewpoint, most simulations work *only* at the price of simplifying the properties of micro-agents, the rules of interaction and the nature of macro-structures so that they conveniently fit each other". In short, the demands of formalising a working model — regardless of its ability to grapple with qualitative data or, for example, heterogeneity — does seem to necessitate a manner of reduction, and a level of simplification that is antithetical to the aim of qualitative research. Thus, whilst it is conceded that simplification and abstraction are entailed in all research endeavours and that ABM — in particular qualitatively driven ABM — holds promise in terms of avoiding the level of abstraction demanded of other complex models (Squazzoni et al. 2014), it is not unproblematic in this respect.

Given this discussion, there does seem to be a danger that in the effort to translate qualitative evidence into something that can be used in a model that evidence is reduced to this entirely. That is, reduced to something which can be formalised by the modeller, and put to work within a model. Whilst adopting this strategy may add to the realism of an ABM, it may well miss the point of much qualitative research, and the commitment of such an approach to thick, contextualised description, which is attentive to messy, and everyday realities. Whether a researcher can be satisfied with this level of abstraction is probably down to the individual him or herself. Whether this is the best way to answer questions about the world is certainly up for debate. In Anna Tsing's (2012, p. 141) words: "there are big stories to be told here" and *attempting* to understand and tell these requires a number of different strategies and methods.[12] In this respect, both ABM and qualitative methods, whilst not incompatible are distinct, each with their own trade-offs and representational capacity. Which raises the question as to how best these stories might be told? Whether they can be told alongside these models? Or whether these models will (further) subsume these?[13]

---

[12] The term stories is used here to denote, for example, rich and varied narrative accounts about how the world unfolds, that go beyond-the-human. Certainly, beyond the micro-level and the individual.

[13] Here, the consequentiality of hyper-simulation versus hyper-quantification comes to mind.

Colonisation, Collaboration and Depictions Otherwise

Thinking in terms of collaborative versus colonising research endeavours, and considering what "collective intellectual experimentation" (Lorimer 2017, p. 133) might look within this space merits reflection on whether there is room for adopting a real commitment to embracing a plurality of approaches. As indicated, using qualitative data to inform models does seem reasonable, mere synthesis decidedly less so (Klenk and Meehan 2015). Given their capacity to grapple with the world in quite different ways, both ABM and qualitative methods have a certain "added-value" (Squazzoni 2010, p. 21). For example, ABM does have the potential to increase our understanding of "reality" in a manner that might not be readily accessible through direct observation, without the aid of a simulation (Frigg and Reiss 2009; Venturini et al. 2015). On the other side of the coin, there are some ways of knowing the world that model-based formalisation simply eludes, and to this end, there are an array of qualitative approaches available that provide a good route to more critical situated engagements that are not so easily rendered amenable to simulation. Consequently, combining these approaches in a manner that resists descending into a narrow integrating, triangulating, or formalising exercise may well offer a promising route towards generating a deeper, more nuanced, careful understanding (Flick 2004), through the incorporation of "multiple lines of sight" (Berg 2001).

Realising this, however, requires a commitment to proceeding in a manner that is mindful that not all knowledge seeks formalisation, nor is it waiting to be *rendered scientific*. Taking Edmonds' (2015) quote from the beginning of this section again – yes qualitative evidence is often criticised for being subjective and context-specific. Arguably, however, there is a lot of room for contextually situated knowledge that is attentive to time, space, politics, economics and culture, and can provide us with less formal, and (perhaps) more critical insights into lived experiences and phenomena. The challenges facing society today are not merely scientific, but rather in many instances political, and ethical. Thus, in these instances, the "added value" of "subjectivism and narrativism" (Squazzoni 2010, p. 21) becomes apparent, whilst that of formalisation is somewhat diminished. In terms of this discussion, it is acknowledged that using qualitative data in an effort to empirically embed models does not represent a conscious attempt at colonisation. Nor do ABMs appear to proclaim a capacity to represent reality in totality (Kwa 1994). Albeit unintended, however, it is suggested that this is potentially what is at stake here, particularly when one considers the wider "intellectual climate" (Castree et al. 2014), and politics of evidence (Denzin 2017; Maxwell 2004), alongside the primacy of the role computer models have hitherto been given in representing the *Anthropocene* (Edwards 1999, 2017).

Conclusion

I began this chapter with a brief discussion of the (dreaded) *Anthropocene*, with the intention of highlighting that this moment raises numerous questions with respect to the politics of problem framing and knowledge production. As detailed, within the current context, much of the emphasis today is trained on the level of the individual, while the broader socio-natural configuration within which this behaviour is structured remains outside of the frame, both within policy circles and a great deal of science. Within this space, rather than a radical reconceptualization of the manner in which we design nature, a very particular currency of social science has gained traction. This backdrop provided the impetus for querying the representational economy of ABMs, including those that are qualitatively driven, and considering what simulating the Anthropocene via ABM looks like. Doing so seems especially pertinent given that complex computer simulations, including ABMs, are gaining increasing precedence across an array of fields related to the socio-environment.

In this respect, ABM does raise a number of philosophical and sociological issues. I have raised only two here, and I have done so with the intention of suggesting that although the world depicted by many ABMs does appear to fit well with the broader narrative, it fits less well with many strands of existing social theory. Considering this, this paper has sought to reinvigorate a long-standing discussion with respect to the ontological base of ABM, and strove to suggest that there is more to this moment than "making up people" (Hacking 1990, pp. 2–3).[14] The discussion raises questions with respect to the dangers of unquestioned assumptions, the *situated-ness* of all knowledge (scientific or other), and excessive formalisation. In this respect, it seems reasonable to suggest that model design should go hand in hand with some serious thought about the manner in which social reality is being represented via ABM, and proceed in a manner that is mindful of the "link between epistemology and social order" (Latour 1993, p. 27), or just ontologies.

Despite these critiques, however, the overarching call here is one of caution rather than outright criticism. Much of the cause for this caution may be related to broader discussions in relation to collaborative, disciplinary hopping (and blurring) research, and how this type of research looks, and might look. Many of the complex challenges on which researchers are focused today *do* require a plurality of views. As such, *any* attempt to answering these, or corresponding calls for more flexible disciplinary approaches, should not be reduced to a narrow assimilation project – methodological or otherwise. Unfortunately, and as indicated in the introduction of this paper, from a certain vantage point, much collaborative research today displays a polemic, flabby and even defensive character. Perhaps this paper might be read

---

[14]The term "making up people" is borrowed from Hacking (1990), whereby he details how, in the nineteenth century, with the enumeration of people and their habits that accompanied the "new technologies for classifying and enumerating" (e.g. statistics), categories were invented into which people could be placed in order to be counted and classified, and by which people subsequently "came to recognize themselves", and others (Hacking 1990, pp. 2–6).

tendentiously as a manifestation of the latter. In acknowledgement, it is clear that a certain level of pragmatism is needed on all sides of this divide. Thus, the call made here is for an *engaged fallibilistic pluralism,* whereby we take our own fallibilities seriously, and make a real attempt to listening to others "without denying or suppressing the otherness of the other" or assuming that we can "always easily translate what is alien into our own entrenched vocabularies" (Bernstein 1989, p. 15).

Acknowledgements Many thanks to Bruce Edmonds, and Ruth Meyer for reading, discussing, and encouraging numerous versions of the argument made here. Thanks also, to Karl Benediktsson for his insightful comments with respect to a final draft, and to Dean Bavington for an introduction to the conceptual affordances of political ontology. I am further grateful for the comments I received from an anonymous reviewer on an earlier version of this paper, and for the feedback from various participants in the Qual2Rule session on an iteration that was presented at the Social Simulation Conference in Dublin in 2017.

Funding This work was funded by the project SAF21 – Social Science Aspects of Fisheries for the 21st Century (project financed under the EU Horizon 2020 Marie Sklodowska-Curie (MSC) ITN-ETN Program; project number: 642080).

References

Bavington, D. (2011). Environmental history during the Anthropocene: Critical reflections on the pursuit of policy-oriented history in the man-age 1. *EH+*. http://niche-canada.org/wp-content/uploads/2014/02/Bavington-commissioned-paper.pdf. Accessed 30 July 2018.

Berg, B. L. (2001). *Qualitative research methods for the social sciences.* Boston: Pearson Education. https://doi.org/10.2307/1317652.

Bernstein, R. J. (1989). Pragmatism, pluralism and the healing of wounds. *Proceedings and Addresses of the American Philosophical Association, 63*(3), 5–18. http://www.jstor.org/stable/3130079. Accessed 18 July 2017.

Bharwani, S., Besa, M. C., Taylor, R., Fischer, M., Devisscher, T., & Kenfack, C. (2015). Identifying salient drivers of livelihood decision-making in the forest communities of Cameroon: Adding value to social simulation models. *Journal of Artificial Societies and Social Simulation, 18*(1). https://doi.org/10.18564/jasss.2646.

Blaser, M. (2009). The threat of the Yrmo: The political ontology of a sustainable hunting program. *American Anthropologist, 111*(1), 10–20. https://doi.org/10.1111/j.1548-1433.2009.01073.x.

Blaser, M. (2013). Ontological conflicts and the stories of peoples in spite of Europe. *Current Anthropology, 54*(5), 547–568. https://doi.org/10.1086/672270.

Boero, R., & Squazzoni, F. (2005). Does empirical embeddedness matter? Methodological issues on agent-based models for analytical social science. *Journal of Artificial Societies and Social Simulation, 8,* 4. http://jasss.soc.surrey.ac.uk/8/4/6.html. Accessed 30 July 2018.

Bourdieu, P. (1989). Social space and symbolic power. *Sociological Theory, 7*(1), 14. https://doi.org/10.2307/202060.

Bulle, N., & Phan, D. (2017). Can analytical sociology do without methodological individualism? *Philosophy of the Social Sciences, 47*(6), 379–409. https://doi.org/10.1177/0048393117713982.

Callon, M. (1984). Some elements of a sociology of translation: Domestication of the scallops and the fishermen of St Brieuc Bay. *The Sociological Review, 32*(S1), 196–233. https://doi.org/10.1111/j.1467-954X.1984.tb00113.x.

Castree, N. (2015). Geography and global change science: Relationships necessary, absent, and possible. *Geographical Research, 53*(1), 1–15. https://doi.org/10.1111/1745-5871.12100.

Castree, N., Adams, W. M., Barry, J., Brockington, D., Büscher, B., Corbera, E., et al. (2014). Changing the intellectual climate. *Nature Climate Change, 4*(9), 763–768. https://doi.org/10.1038/nclimate2339.

Conte, R., Edmonds, B., Moss, S., & Sawyer, R. K. (2001). Sociology and social theory in agent based social simulation: A symposium. *Computational & Mathematical Organization Theory, 7*(3), 183–205. https://doi.org/10.1023/A:1012919018402.

de Castro, E. V. (2015). Who is afraid of the ontological wolf? Some comments on an ongoing anthropological debate. *The Cambridge Journal of Anthropology, 33*(1), 2–17. https://doi.org/10.3167/ca.2015.330102.

Denzin, N. K. (2017). Critical qualitative inquiry. *Qualitative Inquiry, 23*(1), 8–16. https://doi.org/10.1177/1077800416681864.

Edmonds, B. (2015). Using qualitative evidence to inform the specification of agent-based models. *Journal of Artificial Societies and Social Simulation, 18*(1). https://doi.org/10.18564/jasss.2762.

Edwards, P. N. (1999). Global climate science, uncertainty and politics: Data-laden models, model-filtered data. *Science as culture, 8*(4), 437–472.

Edwards, P. N. (2017). Knowledge infrastructures for the Anthropocene. *The Anthropocene Review, 4*(1), 34–43.

Elder-Vass, D. (2017). Material parts in social structures. *Journal of Social Ontology, 3*(1), 89–105. https://doi.org/10.1515/jso-2015-0058.

Emirbayer, M. (1997). Manifesto for a relational sociology. *American Journal of Sociology, 103*(2), 281–317. https://doi.org/10.1086/231209.

Epstein, B. (2013). Agent-based modeling and the fallacies of individualism. In *Models simulations and representations* (pp. 115–144). New York: Routledge. https://doi.org/10.4324/9780203808412.

Epstein, B. (2016a). A framework for social ontology. *Philosophy of the Social Sciences, 46*(2), 147–167. https://doi.org/10.1177/0048393115613494.

Epstein, B. (2016b). Précis of the ant trap. *Journal of Social Ontology, 2*(1), 125–134. https://doi.org/10.1515/jso-2016-0001.

Feyerabend, P. (2001). *Conquest of abundance: A tale of abstraction versus the richness of being.* Chicago/London: University of Chicago Press. SUNY Fredonia, B3240 F483 C66 1999.

Flick, U. (2004). Triangulation in qualitative resarch. In I. Kardoff & E. Steinke (Eds.), *A companion to qualitative research.* London: Sage. http://www.sdhprc.ir/download/A_Companion_to_qualitative_research.pdf#page=193. Accessed 10 August 2018.

Frigg, R., & Reiss, J. (2009). The philosophy of simulation: Hot new issues or same old stew? *Synthese, 169*, 593–613. https://doi.org/10.1007/s11229-008-9438-z.

Government Office for Science. (2018). *Computational modelling: Technological futures.* https://assets.publishing.service.gov.uk/government/uploads/system/uploads/attachment_data/file/682579/computational-modelling-blackett-review.pdf. Accessed 30 July 2018.

Granovetter, M. (1985). Economic action and social structure: The problem of embeddedness'. *American Journal of Sociology, 91*(3), 481–510. https://doi.org/10.2307/2780199.

Hacking, I. (1990). *The taming of chance.* Cambridge: Cambridge University Press.

Haraway, D. (1988). Situated knowledges: The science question in feminism and the privilege of partial perspective. *Feminist Studies, 14*(3), 575. https://doi.org/10.2307/3178066.

Hay, C. (2013). *Political ontology Oxford handbooks online.* https://doi.org/10.1093/oxfordhb/9780199604456.013.0023.

Hedström, P., & Ylikoski, P. (2010). Causal mechanisms in the social sciences. *Annual Review of Sociology, 36*(1), 49–67. https://doi.org/10.1146/annurev.soc.012809.102632.

Holbraad, M., Pedersen, M. A., & de Castro, E. (2014). The politics of ontology: Anthropological positions. *Cultural Antrhopology,* (13), 365–387. https://culanth.org/fieldsights/462-the-politics-of-ontology-anthropological-position. Accessed 10 August 2018.

Klenk, N., & Meehan, K. (2015). Climate change and transdisciplinary science: Problematizing the integration imperative. *Environmental Science and Policy, 54*, 160–167. https://doi.org/10.1016/j.envsci.2015.05.017.

Knorr-Cetina, K. D. (1988). The micro-social order: Towards a reconception. In N. Fielding (Ed.), *Actions and structure: Research methods and social theory* (pp. 21–53). London: Sage.

Kohn, E. (2015). Anthropology of ontologies. *Annual Review of Anthropology, 44*(1), 311–327. https://doi.org/10.1146/annurev-anthro-102214-014127.

Kwa, C. (1994). Modelling technologies of control. *Science as Culture, 4*(3), 363–391. https://doi.org/10.1080/09505439409526393.

Latour, B. (1993). *We have never been modern*. Cambridge: Harvard University Press.

Little, D. (2012). Analytical sociology and the rest of sociology. *Sociologica, 6*(1), 1–47. https://doi.org/10.2383/36894.

Long, N., & Long, A. (1992). *Battlefields of knowledge: The interlocking of theory and practice in social research and development*. London/New York: Routledge.

Lorimer, J. (2017). The Anthropo-scene: A guide for the perplexed. *Social Studies of Science, 47*(1), 117–142. https://doi.org/10.1177/0306312716671039.

Lövbrand, E., Beck, S., Chilvers, J., Forsyth, T., Hedrén, J., Hulme, M., et al. (2015). Who speaks for the future of Earth? How critical social science can extend the conversation on the Anthropocene. *Global Environmental Change, 32*, 211–218. https://doi.org/10.1016/j.gloenvcha.2015.03.012.

MacLure, M. (2017). Qualitative methodology and the new materialisms: "a little of Dionysus's blood?". In M. Denzin & N. K. Giardina (Eds.), *Qualitative inquiry in neoliberal times* (pp. 48–58). New York: Routledge. https://doi.org/10.4324/9781315397788-14.

Marchionni, C., & Ylikoski, P. (2013). Agent-based simulation. *Philosophy of the Social Sciences, 43*(3), 323–340. https://doi.org/10.1177/0048393113488873.

Maxwell, J. A. (2004). Reemergent scientism, postmodernism, and dialogue across differences. *Qualitative Inquiry, 10*(1), 35–41. https://doi.org/10.1177/1077800403259492.

Mitchell, T. (2002). *Rule of experts: Egypt, techno-politics, modernity*. Berkeley/London: University of California Press. https://www.ucpress.edu/book/9780520232624/rule-of-experts. Accessed 31 July 2018.

Mol, A. (1999). Ontological politics. A word and some questions. *The Sociological Review, 47*(1_suppl), 74–89. https://doi.org/10.1111/j.1467-954X.1999.tb03483.x.

Moore, J. W. (2016). Introduction. Anthropocene or capitalocene? Nature, history and the crisis of capitalism. In J. W. Moore (Ed.), *Anthropocene or capitalocene? Nature, history and the crisis of capitalism* (pp. 1–13). Oakland: Pm Press. https://doi.org/10.1017/CBO9781107415324.004.

Morton, T. (2013). *Hyperobjects: Philosophy and ecology after the end of the world*. London: University of Minnesota Press. https://doi.org/10.1017/CBO9781107415324.004.

Morton, T. (2017). *Humankind: Solidarity with non-human people*. London: Verso Books.

O'Sullivan, D., & Haklay, M. (2000). Agent-based models and individualism: Is the world agent-based? *Environment and Planning A, 32*(8), 1409–1425. https://doi.org/10.1068/a32140.

Pellizzoni, L. (2015). *Ontological politics in a disposable world: The new mastery of nature. Science as culture* (Vol. 25). New York: Routledge. https://doi.org/10.1080/09505431.2016.1172562.

Pilkey, O. H., & Pilkey-Jarvis, L. (2007). *Useless arithmetic: Why environmental scientists can't predict the future*. New York: Columbia University Press. https://doi.org/10.1080/00330120701724426.

Polanyi, K. (1944). *The great transformation: The political and economic origins of our time*. New York: Rinehart.

Ranciere, J. (2001). Ten theses on politics. *Theory & Event, 5*(3), 1–16. https://doi.org/10.1353/tae.2001.0028.

Robbins, P., & Moore, S. A. (2013). Ecological anxiety disorder: Diagnosing the politics of the Anthropocene. *Cultural Geographies, 20*(1), 3–19. https://doi.org/10.1177/1474474012469887.

Rose, G. (1997). Situating knowledges: Positionality, reflexivities and other tactics. *Progress in Human Geography, 21*(3), 305–320. https://doi.org/10.1191/030913297673302122.

Saltelli, A., & Giampietro, M. (2017). What is wrong with evidence based policy, and how can it be improved? *Futures, 91*, 62–71. https://doi.org/10.1016/J.FUTURES.2016.11.012.

Sawyer, K. R. (2005). *Social emergence: Societes as complex systems*. Cambridge: Cambridge University Press. http://www.cambridge.org/ie/academic/subjects/sociology/social-theory/social-emergence-societies-complex-systems?format=PB&isbn=9780521606370. Accessed 31 July 2018.

Squazzoni, F. (2010). The impact of agent-based models in the social sciences after 15 years of incursions. *History of Economic Ideas, 18*(2), 197–234. https://econpapers.repec.org/article/hidjournl/v_3a18_3ay_3a2010_3a2_3a8_3ap_3a197-234.htm. Accessed 31 July 2018.

Squazzoni, F., Jager, W., & Edmonds, B. (2014). Social simulation in the social sciences. *Social Science Computer Review, 32*(3), 279–294. https://doi.org/10.1177/0894439313512975.

Swyngedouw, E., & Ernstson, H. (2018). Interrupting the Anthropo-obScene: Immuno-biopolitics and depoliticizing ontologies in the Anthropocene. *Theory, Culture and Society, 35*(6), 3–30. https://doi.org/10.1177/0263276418757314.

Tsing, A. (2012). Unruly edges: Mushrooms as companion species. *Environmental Humanities, 1*(1), 141–154. https://doi.org/10.1215/22011919-3610012.

Venturini, T., Jensen, P., & Latour, B. (2015). Fill in the gap. A new alliance for social and natural sciences. *Journal of Artificial Societies and Social Simulation, 18*(2), 1–4. https://doi.org/10.18564/jasss.2729.

Yang, L., & Gilbert, N. (2008). Getting away from numbers: Using qualitative observation for agent-based modeling. *Advances in Complex Systems, 11*(02), 175–185. https://doi.org/10.1142/S0219525908001556.

Žižek, S. (2017). *The courage of hopelessness. Chronicles of a year of acting dangerously*. London: Allen Lane.

Chapter 3
The Art of Iterating: Update-Strategies in Agent-Based Simulation

Jonathan Thaler and Peer-Olaf Siebers

Introduction

Agent-based simulation (ABS) is a method for simulating the emergent behaviour of a system by modelling and simulating the interactions of its subparts, called agents. Examples for an ABS are simulating the spread of an epidemic throughout a population or simulating the dynamics of segregation within a city. Central to ABS is the concept of an agent who needs to be updated in regular intervals during the simulation so it can interact with other agents and its environment. In this paper we are looking at two different kind of simulations to show the differences update-strategies can make in ABS and support our main message that when developing a model for an ABS it is of most importance to select the right update-strategy which reflects and supports the corresponding semantics of the model. As we will show, due to conflicting ideas about update-strategies, this awareness is yet still underrepresented in the field of ABS and is lacking a systematic treatment. As a remedy we undertake such a systematic treatment in proposing a new terminology by identifying properties of ABS and deriving all possible update-strategies. The outcome is a terminology to communicate in a unified way this very important matter, so researchers and implementers can talk about it in a common way, enabling better discussions, same understanding and better reproducibility and continuity in research. The two simulations we use are discrete and continuous games where in the former one the agents act synchronized at discrete time-steps whereas in the later one they act continuously in continuous time. We show that in the case of simulating the discrete game the update-strategies have a huge impact on the final result whereas our continuous game seems to be stable under different update-strategies. The contribution of this paper is: Identifying general properties of ABS, deriving update-strategies from these properties and establishing a general terminology for talking about these update-strategies.

J. Thaler (✉) · P.-O. Siebers
School of Computer Science, University of Nottingham, Nottingham, UK
e-mail: jonathan.thaler@nottingham.ac.uk; peer-olaf.siebers@nottingham.ac.uk

© Springer Nature Switzerland AG 2019
D. Payne et al. (eds.), *Social Simulation for a Digital Society*, Springer
Proceedings in Complexity, https://doi.org/10.1007/978-3-030-30298-6_3

Background

In this section we define our understanding of agent and ABS and how we understand and use it in this paper. Then we will give a description of the two kinds of games which motivated our research and case-studies. Finally we will present related work.

Agent-Based Simulation

We understand ABS as a method of modelling and simulating a system where the global behaviour may be unknown but the behaviour and interactions of the parts making up the system is known. Those parts, called agents, are modelled and simulated out of which then the aggregate global behaviour of the whole system emerges. So the central aspect of ABS is the concept of an agent which can be understood as a metaphor for a pro-active unit, situated in an environment, able to spawn new agents and interacting with other agents in a network of neighbours by exchange of messages (Wooldridge 2009). It is important to note that we focus our understanding of ABS on a very specific kind of agents where the focus is on communicating entities with individual, localized behaviour from out of which the global behaviour of the system emerges. We informally assume the following about our agents:

- They are uniquely addressable entities with some internal state.
- They can initiate actions on their own e.g. change their internal state, send messages, create new agents, kill themselves.
- They can react to messages they receive with actions as above.
- They can interact with an environment they are situated in.

An implementation of an ABS must solve two fundamental problems: (1) How can an agent initiate actions without the external stimuli of messages? (source of pro-activity) and (2) When is a message m, sent by agent A to agent B, visible and processed by B? (semantics of messaging). In computer systems, pro-activity, the ability to initiate actions on its own without external stimuli, is only possible when there is some internal stimulus, most naturally represented by a continuous increasing time-flow. Due to the discrete nature of a computer-system, this time-flow must be discretized in steps as well and each step must be made available to the agent, acting as the internal stimulus. This allows the agent then to perceive time and become pro-active depending on time. So we can understand an ABS as a discrete time-simulation where time is broken down into continuous (using real numbers) or discrete (using natural numbers) time-steps. Independent of the representation of the time-flow we have the two fundamental choices whether the time-flow is local to the agent or whether it is a system-global time-flow. Time-flows in computer-systems can only be created through threads of execution where there are two ways of feeding time-flow into an agent. Either it has its own thread-of-execution or the system creates the illusion of its own thread-of-execution by sharing the global thread sequentially among the agents where an agent has to yield the execution back after it has executed its step. Note the similarity to an operating system with cooperative multitasking in the latter case and real multi-processing in the former.

The semantics of messaging define when sent messages are visible to the receivers and when the receivers process them. Message-processing could happen either immediately or delayed, depending on how message-delivery works. There are two ways of message-delivery: immediate or queued. In the case of immediate message-deliver the message is sent directly to the agent without any queuing in between e.g. a direct method-call. This would allow an agent to immediately react to this message as this call of the method transfers the thread-of-execution to the agent. This is not the case in the queued message-delivery where messages are posted to the message-box of an agent and the agent pro-actively processes the message-box at regular points in time.

A Discrete Game: Prisoners Dilemma

As an example of a discrete game we use the Prisoners Dilemma as presented in Nowak and May (1992). In the prisoners dilemma one assumes that two persons are locked up in a prison and can choose to cooperate with each other or to defect by betraying the other one. Looking at a game-theoretic approach there are two options for each player which makes four possible outcomes. Each outcome is associated with a different payoff in the prisoner-dilemma. If both players cooperate both receive payoff R; if one player defects and the other cooperates the defector receives payoff T and the cooperator payoff S; if both defect both receive payoff P where $T > R > P > S$. The dilemma is that the safest strategy for an individual is to defect but the best payoff is only achieved when both cooperate. In the version of Nowak and May (1992) $N \times N$ agents are arranged on a 2D-grid where every agent has eight neighbours except at the edges. Agents do not have a memory of the past and have one of two roles: either cooperator or defector. In every step an agent plays the game with all its neighbours, including itself and sums up the payoff. After the payoff sum is calculated the agent changes its role to the role of the agent with the highest payoff within its neighbourhood (including itself). The authors attribute the following payoffs: $S = P = 0$, $R = 1$, $T > b$, where $b > 1$. They showed that when having a grid of only cooperators with a single defector at the center, the simulation will form beautiful structural patterns as shown in Fig. 3.1.

In Huberman and Glance (1993) the authors show that the results of simulating the Prisoners Dilemma as above depends on a very specific strategy of iterating the simulation and show that the beautiful patterns seen in Fig. 3.1 will not form when selecting a different update-strategy. They introduced the terms of synchronous and asynchronous updates and define synchronous to be as agents being updated in unison and asynchronous where one agent is updated and the others are held

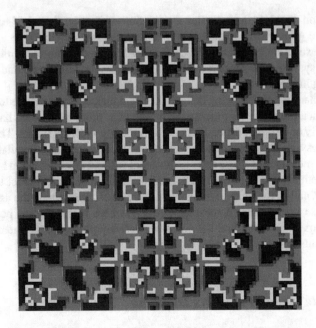

Fig. 3.1 Patterns formed by playing the Prisoners Dilemma game. Notes: Patterns formed by playing the Prisoners Dilemma game on a 99×99 grid with $1.8 < b < 2$ after 217 steps with all agents being cooperators except one defector at the center. Blue are cooperators, red are defectors, yellow are cooperators which were defectors in the previous step, green are defectors which were cooperators in the previous step. (Picture taken from Nowak and May (1992))

constant. Only the synchronous updates are able to reproduce the results. The authors differentiated between the two strategies but their description still lacks precision and detail, something we will provide in this paper. Although they published their work in the area on general computing, it has implications for ABS as well which can be generalized in the main message of our paper as emphasised in the introduction. We will show that there are more than two update-strategies and will give results of simulating this discrete game using all of them. As will be shown later, the patterns emerge indeed only when selecting a specific update-strategy.

A Continuous Game: Heroes & Cowards

As an example for a continuous game we use the Heroes & Cowards game introduced by Wilensky and Rand (2015). In this game one starts with a crowd of agents where each agent is positioned randomly in a continuous 2D-space which is bounded by borders on all sides. Each of the agents then selects randomly one friend and one enemy (except itself) and decides randomly whether the agent acts in the role of a hero or a coward – friend, enemy and role do not change after the initial set-up. In

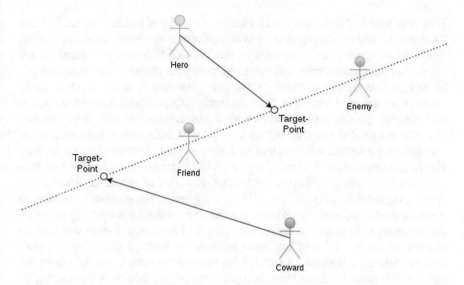

Fig. 3.2 A conceptual diagram of the Heroes & Cowards game. Note: Hero (green) and coward (red) have the same agents as friend and enemy but act different: the hero tries to move in between the friend and enemy whereas the coward tries to hide behind its friend

each step the agent will move a small distance towards a target point as seen in Fig. 3.2. If the agent is in the role of a hero this target point will be the half-way distance between the agent's friend and enemy – the agent tries to protect the friend from the enemy. If the agent is acting like a coward it will try to hide behind the friend also the half-way distance between the agent's friend and enemy, just in the opposite direction. Note that this simulation is determined by the random starting positions, random friend and enemy selection, random role selection and number of agents. Note also that during the simulation-stepping no randomness is incurred and given the initial random set-up, the simulation-model is completely deterministic. As will be shown later the results of simulating this model are invariant under different update-strategies.

Related Research

Besides Nowak and May (1992) and Huberman and Glance (1993) which both discuss asynchronous and synchronous updates, multiple other works mention these kind of updates but the meaning is different in each. Asynchronous updates in the context of cellular automata was defined by Bersini and Detours (1994) as picking a cell at random and updating it and synchronous as all cells updating at the same time and report different dynamics when switching between the two. The authors also raise the question which of both is correct and most faithful to reality as in an ideal solution both should deliver the similar spatio-temporal dynamics.

They conclude that developers of simulations should pay attention to the fact that the dynamics and results are sensible to the updating procedures. Asynchronous vs. synchronous updates are mentioned in the book of Wilensky and Rand (2015) where they define asynchronous updates as having the property that changes made by an agent are seen immediately by the others whereas in synchronous updating the changes are only visible in the next tick. They also look into the notion of sequential vs. parallel actions and identify as sequential when only one agent acts at a time and parallel when agents act truly parallel, independent from each other. The same argumentation is followed by Railsback and Grimm (2011) where they discuss the importance of order of execution of the agents and describe asynchronous and synchronous updating. Yet another definition of synchronous and asynchronous updates is given in Page (1997). They define asynchronous updating as updating agents sequentially one after another and synchronous updating as updating all agents at virtually the same time. They go further and discuss also random updates where the order of the agent-sequence is shuffled before updating all agents. They also introduce incentive based asynchronous updating where the agent which gains the most from the update is updated first, thus introducing an ordering on the sequence which is sorted by the benefit each agent gains from its update. They also compare the differences synchronous, random-asynchronous and incentive-asynchronous updating has on dynamics and come to the conclusion that the order of updating the agents has an impact on the dynamics and should be considered with great care when implementing a simulation. Asynchronous and synchronous time-models are mentioned in Dawson et al. (2014) where the authors describe basic inner workings of ABS environments and compare their implementation in C++ to the existing ABS environment AnyLogic which is programmed in Java. They interpret asynchronous time-models to be the ones in which an agent acts at random time intervals and synchronous time-models where agents are updated all in same time intervals. A different interpretation of synchronous and asynchronous time-models is given in Yuxuan (2016). He identifies the asynchronous time-model to be one in which updates are triggered by the exchange of messages and the synchronous ones which trigger changes immediately without the indirection of messages.

A different approach was taken in Botta et al. (2010) where they sketch a minimal ABS implementation in Haskell. Their research applies primarily to economic simulations and instead of iterating a simulation with a global time, their focus is on how to synchronize agents which have internal, local transition times. A very different approach to updating and iterating agents in ABS than to mechanisms used in existing software like AnyLogic or NetLogo was given in Lysenko et al. (2008) where the authors mapped ABS on Graphics Processing Units (GPU). They discuss execution order at length, highlight the problem of inducing a specific execution-order in a model which is problematic for parallel execution and give solutions how to circumvent these shortcomings. Although we haven't mapped our ideas to GPUs we explicitly include an approach for data-parallelism which can be utilized to roughly map their approach onto our terminology.

A New Terminology

When looking at related work, we observe that there seems to be a variety of meanings attributed to the terminology of asynchronous and synchronous updates, but the very semantic and technical details are unclear and not described very precisely. To develop a standard terminology, we propose to abandon the notion of synchronous and asynchronous updates and, based on the discussion above, we propose six properties characterizing the dimensions and details of the internals of an ABS. Having these properties identified, we then derive all meaningful and reasonable update-strategies which are possible in a general form in ABS. These update-strategies together with the properties will form the new terminology we propose for speaking about update-strategies in ABS in general. We will discuss all details programming-language agnostic and for each strategy we give a short description, the list of all properties and discuss their semantics, variations and implications selecting update-strategies for a model. A summary of all update-strategies and their properties is given in Table 3.1.

ABS Properties

We identified the following properties of agent-based simulations, which are necessary to derive and define the differences between the update-strategies.

Iteration-order: Is the collection of agents updated *sequentially* with one agent updated after the other or are all agents updated in *parallel*, at virtually the same time?

Global synchronization: Is a full iteration over the collection of agents happening in lock-step at global points in time or not *(yes/no)*?

Thread of execution: Does each agent have a *separate* thread of execution or does it *share* it with all others? Note that it seems to add a constraint on the Iteration-Order, namely that *parallel* execution forces separate threads of execution for all agents. We will show that this is not the case, when looking at the parallel strategy in the next section.

Table 3.1 Update-strategies in ABS

	Sequential	Parallel	Concurrent	Actor
Iteration-order	Sequential	Parallel	Parallel	Parallel
Global-sync	Yes	Yes	Yes	No
Thread	Shared	Separate	Separate	Separate
Messaging	Immediate	Queued	Queued	Queued
Visibility	In	Post	In	In
Repeatability	Yes	Yes	No	No

Message-handling: Are messages handled *immediately* by an agent when sent to them or are they *queued* and processed later? Here we have the constraint that an immediate reaction to messages is only possible when the agents share a common thread of execution. Note that we must enforce this constraint as otherwise agents could end up having more than one thread of execution which could result in them acting concurrently by making simultaneous actions. This is something we explicitly forbid as it is against our definition of agents which allows them to have only one thread of execution at a time.

Visibility of changes: Are the changes made (messages sent, environment modified) by an agent which is updated during an Iteration-Order visible (during) *In-Iteration* or only *Post-Iteration* at the next Iteration-Order? More formally: do agents $a_{n>1}$, which are updated after agent a_i, see the changes by agent a_i or not? If yes, we refer to *In-Iteration* visibility, to *Post-Iteration* otherwise.

Repeatability: Does the ABS have an external source of non-determinism which it cannot influence? If this is the case, then we regard an update-strategy as *non-deterministic*, otherwise *deterministic*. It is important to distinguish between *external* and *internal* sources of non-determinism. The former are race-conditions due to concurrency, creating non-deterministic orderings of events which has the consequence that repeated runs may lead to different results with the same configuration, rendering an ABS non-deterministic. The latter, coming from random-number generators, can be controlled using the same starting-seed leading to repeatability and deemed deterministic in this context.

ABS Update-Strategy: Sequential

This strategy has a globally synchronized time-flow and in each time-step iterates through all the agents and updates one agent after another. Messages sent and changes to the environment made by agents are visible immediately.

Iteration-order: Sequential
Global synchronization: Yes
Thread of execution: Shared
Message-handling: Immediate (or Queued)
Visibility of changes: In-Iteration
Repeatability: Deterministic

Semantics: There is no source of randomness and non-determinism, rendering this strategy to be completely deterministic in each step. Messages can be processed either immediately or queued depending on the semantics of the model. If the model requires to process the messages immediately the model must be free of potential infinite loops.

Variation: If the sequential iteration from agent [1 ... n] imposes an advantage over the agents further ahead or behind in the queue (e.g. if it is of benefit when making choices earlier than others in auctions or later when more information is available) then one could use random-walk iteration where in each time-step the agents are shuffled before iterated. Note that although this would introduce randomness in the model the source is a random-number generator implying it is still deterministic. If one wants to have a very specific ordering, e.g. 'better performing' agents first, then this can be easily implemented too by exposing some sorting-criterion and sorting the collection of agents after each iteration.

ABS Update-Strategy: Parallel

This strategy has a globally synchronized time-flow and in each time-step iterates through all the agents and updates them in parallel. Messages sent and changes to the environment made by agents are visible in the next global step. We can think about this strategy in a way that all agents make their moves at the same time.

Iteration-order: Parallel
Global synchronization: Yes
Thread of execution: Separate (or Shared)
Message-handling: Queued
Visibility of changes: Post-Iteration
Repeatability: Deterministic

Semantics: If one wants to change the environment in a way that it would be visible to other agents this is regarded as a systematic error in this strategy. First it is not logical because all actions are meant to happen at the same time and also it would implicitly induce an ordering, violating the happens at the same time idea. To solve this, we require different semantics for accessing the environment in this strategy. We introduce a global environment which is made up of the set of local environments. Each local environment is owned by an agent so there are as many local environments as there are agents. The semantics are then as follows: in each step all agents can read the global environment and read/write their local environment. The changes to a local environment are only visible after the local step and can be fed back into the global environment after the parallel processing of the agents. It does not make a difference if the agents are really computed in parallel or just sequentially – due to the isolation of information, this has the same effect. Also, it will make no difference if we iterate over the agents sequentially or randomly, the outcome has to be the same: the strategy is event-ordering invariant as all events and updates happen virtually at the same time. If one needs to have the semantics of writes on the whole (global) environment in one's model, then this strategy is not the right one and one should resort to one of the other strategies. A workaround would be to implement the global environment as an agent with which the non-environment agents can communicate via messages introducing an ordering but which is then sorted in a controlled way by an agent, something which is not possible in the case of a passive, non-agent environment. It is important to note that in this strategy a reply to a message will not be delivered in the current but in the next global time-step. This is in contrast to the immediate message-delivery of the sequential strategy where within a global time-step agents can have in fact an arbitrary number of messages exchanged.

ABS Update-Strategy: Concurrent

This strategy has a globally synchronized time-flow and in each time-step iterates through all the agents and updates all agents in parallel but all messages sent and changes to the environment are immediately visible. So this strategy can be understood as a more general form of the parallel strategy: all agents run at the same time but act concurrently.

Iteration-order: Parallel
Global synchronization: Yes
Thread of execution: Separate
Message-handling: Queued
Visibility of changes: In-Iteration
Repeatability: Non-Deterministic

Semantics: It is important to realize that when running agents in parallel who are able to see actions by others immediately this is the very definition of concurrency: parallel execution with mutual read/write access to shared data. Of course this shared data-access needs to be synchronized which in turn will introduce event-orderings in the execution of the agents. At this point we have a source of inherent non-determinism: although when one ignores any hardware-model of concurrency, at some point we need arbitration to decide which agent gets access first to a shared resource arriving at non-deterministic solutions. This has the very important consequence that repeated runs with the same configuration of the agents and the model may lead to different results.

ABS Update-Strategy: Actor

This strategy has no globally synchronized time-flow, but all the agents run concurrently in parallel with their own local time-flow. The messages and changes to the environment are visible as soon as the data arrive at the local agents. This can be immediately when running locally on a multi-processor or with a significant delay when running in a cluster over a network. Obviously, this is also a non-deterministic strategy and repeated runs with the same agent- and model-configuration may (and will) lead to different results.

Iteration-order: Parallel
Global synchronization: No
Thread of execution: Separate
Message-handling: Queued
Visibility of changes: In-Iteration
Repeatability: Non-Deterministic

Semantics: It is of importance to note that information and also time in this strategy are always local to an agent as each agent progresses in its own speed through the simulation. In this case, one needs to explicitly observe an agent when one wants to e.g. visualize it. This observation is then only valid for this current point in time, local to the observer, but not to the agent itself, which may have changed immediately after the observation. This implies that we need to sample our agents with observations when wanting to visualize them, which would inherently lead to well-known sampling issues. A solution would be to invert the problem and create an observer-agent which is known to all agents where each agent sends an 'I have changed' message with the necessary information to the observer if it has changed its internal state. This also does not guarantee that the observations will really reflect the actual state the agent is in but is a remedy against the previously mentioned sampling problems. Problems can occur though if the observer-agent cannot process the update-messages fast enough, resulting in a congestion of its message-queue. The concept of Actors was proposed by Hewitt et al. (1973) for which Greif (1975) and Clinger (1981) developed semantics of different kinds. These works were very influential in the development of the concepts of agents and can be regarded as foundational basics for ABS.

Variation: This is the most general one of all the strategies as it can emulate all the others by introducing the necessary synchronization mechanisms.

ABS Toolkits

There exist a lot of tools for modelling and running ABS. We investigated the abilities of two of them to capture our update-strategies and give an overview of our findings in this section.

NetLogo: NetLogo is probably the most popular ABS toolkit around as it comes with a modelling language which is very close to natural language and very easy to learn for non-computer scientists. It follows a strictly single-threaded computing approach when running a single model, so we can rule out both the concurrent and actor strategy as both require separate threads of execution. The tool has no built-in concept of messages and it is built on global synchronization which is happening through advancing the global time by the 'tick' command. It falls into the responsibility of the model-implementer to iterate over all agents and let them perform actions on themselves and on others. This allows for very flexible updating of agents which also allows to implement the parallel strategy. A NetLogo model which implements the prisoners dilemma game synchronous and asynchronous to reproduce the findings of (Huberman and Glance 1993) can be found in Sect. 5.4 of (Jansen 2012).

AnyLogic: AnyLogic follows a rather different approach than NetLogo and is regarded as a multi-method simulation tool as it allows to do system dynamics, discrete event simulation and agent-based simulation at the same time where all three methods can interact with each other. For ABS it provides the modeler with a high-level view on agents and does not provide the ability to iterate over all agents – this is done by AnyLogic itself and the modeler can customize the behaviour of an agent either by modelling diagrams or programming in Java. As NetLogo, AnyLogic runs a model using a single thread, thus, the concurrent and actor strategy are not feasible in AnyLogic. A feature this toolkit provides is communication between agents using messages and it supports both queued and immediate messaging. AnyLogic does not provide a mechanism to directly implement the parallel strategy because all changes are seen immediately by the other agents but using queued messaging we think that the parallel strategy can be emulated nevertheless.

To conclude, the most natural and common update-strategy in these toolkits is the sequential strategy which is not very surprising. The primary target are mostly agent-based modelers who are non-computer scientists so the toolkits also try to be as simple as possible and multi-threading and concurrency would introduce lots of additional complications for modelers to worry about. So the general consensus is to refrain from multi-threading and concurrency as it is obviously harder to develop, debug and introduces non-repeatability in the case of concurrency and to stick with the sequential strategy. The parallel strategy is not supported directly by any of them but can be implemented using various mechanisms like queued message passing and custom iteration over the agents.

Case Studies

In this section we present two case-studies, the Prisoners Dilemma and Heroes & Cowards games, for discussing the effect of different update-strategies. As already emphasized, both are of different nature. The first one is a discrete game, which is played at discrete time-steps. The second one is a continuous game where each agent is continuously playing. This has profound implications on the simulation results shown below. We implemented the simulations for all strategies except the actor strategy in Java and the simulations for the actor strategy in Haskell and in Scala with the Actor-Library.

Prisoners Dilemma

Our agent-based model of this game works as follows: at the start of the simulation each agent sends its state to all its neighbours which allows to incrementally calculate the local payoff. If all neighbours' states have been received then the agent will send its local payoff to all neighbours which allows to compare all payoffs in its neighbourhood and calculate the best. When all neighbours' local payoffs have been received the agent will adopt the role of the highest payoff and sends its new state to all its neighbours, creating a circle. Care must be taken to ensure that the update-strategies are comparable because when implementing a model in an update-strategy it is necessary to both map the model to the strategy and try to stick to the same specification – if the implementation of the model differs fundamentally across the update-strategies it is not possible to compare the solutions. So we put great emphasis and care keeping all four implementations of the model the same just with a different update-strategy running behind the scenes which guarantees comparability.

The results as seen in the left column of Fig. 3.3 were created with the same configuration as reported in Nowak and May (1992). When comparing the pictures with the one from the reference seen in Fig. 3.1, the only update-strategy which is able to reproduce the matching result is the parallel strategy – all the others clearly fail to reproduce the pattern. From this we can tell that only the parallel strategy is suitable to simulate this model.

To reproduce the pattern of Fig. 3.1, the simulation needs to be split into two global steps which must happen after each other: first calculating the sum of all payoffs for every agent and then selecting the role of the highest payoff within the neighbourhood. This two-step approach results in the need for twice as many steps to arrive at the matching pattern when using *queued* messaging as is the case in the parallel, concurrent and actor strategy.

For the sequential strategy one must further differentiate between immediate and queued messaging. We presented the results using the queued version, which has the same implementation of the model as the others. When one is accepting to change the implementation slightly, then the immediate version is able to arrive at the pattern after 217 steps with a slightly different model-implementation: because immediate messaging transfers the thread of control to the receiving agent that agent can reply within this same step. This implies that we can calculate a full game-round (both steps) within one global time-step by a slight change in the model- implementation: an agent sends its current state in every time-step to all its neighbours.

The reason why the other strategies fail to reproduce the pattern is due to the non-parallel and unsynchronized way that information spreads through the grid. In the sequential strategy the agents further ahead in the queue play the game earlier and influence the neighbourhood so agents which play the game later find already messages from earlier agents in their queue, thus acting differently based upon this information. Also, agents will send messages to themselves which will be processed in the same time-step. In the concurrent and actor strategy the agents run in parallel

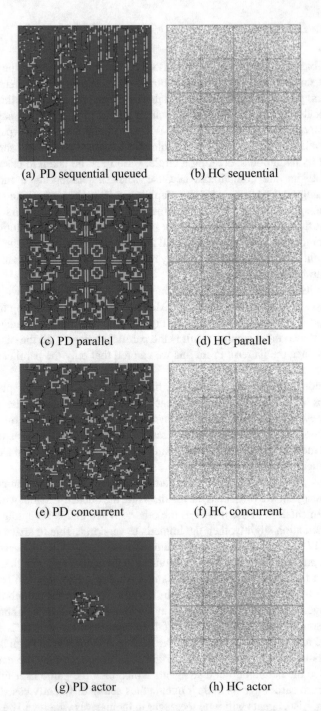

(a) PD sequential queued   (b) HC sequential

(c) PD parallel   (d) HC parallel

(e) PD concurrent   (f) HC concurrent

(g) PD actor   (h) HC actor

Fig. 3.3 Effect on results simulating the Prisoners Dilemma (PD) and Heroes & Cowards (HC) with all four update-strategies

but changes are visible immediately and concurrently, leading to the same non-structural patterns as in the sequential one. Although agents do not change unless all their neighbours have answered, this does not guarantee a synchronized update of all agents because every agent has a different neighbourhood, which is reflexive but not transitive. If agent a is agent's b neighbour and agent c is agent's b neighbour this does not imply that agent c is agent's a neighbour as well. This allows the spreading of changes throughout the neighbourhood, resulting in a breaking down of the pattern. This is not the case in the parallel strategy where all agents play the game at the same time based on the frozen state of the previous step, leading to a synchronized update as required by the model. Note that the concurrent and actor strategy produce different results on every run due to the inherent non-deterministic event-ordering introduced by concurrency.

Heroes & Cowards

Our agent-based model of this game works as follows: in each time-step an agent asks its friend and enemy for their positions which will answer with a corresponding message containing their current positions. The agent will have its own local information about the position of its friend and enemy and will calculate its move in every step based on this local information.

The results as seen in the right column of Fig. 3.3 were created with 100.000 agents where 25% of them are heroes running for 500 steps. Although the individual agent-positions of runs with the same configuration differ between update-strategies the cross-patterns are forming in all four update-strategies. For the patterns to emerge it is important to have significant more cowards than heroes and to have agents in the tens of thousands – we went for 100.000 because then the patterns are really prominent. The patterns form because the heroes try to stay halfway between their friend and enemy: with this high number of cowards it is very likely that heroes end up with two cowards – the cowards will push towards the border as they try to escape, leaving the hero in between. We can conclude that the Heroes & Cowards model seems to be robust to the selection of its update-strategy and that its emergent property – the formation of the cross – is stable under differing strategies.

Conclusion and Future Research

In this paper we have presented general properties of ABS, derived four general update-strategies and discussed their implications. By doing this we proposed a unified terminology which allows to speak about update-strategies in a common and unified way, something that the ABS community is currently lacking. We hope our classification and terminology will help the community to better understand the details necessary to consider implementing an agent-based simulation. Again, we cannot stress enough that selecting the right update-strategy is of most importance and must match the semantics of the model one wants to simulate. We showed that the Prisoners Dilemma game on a 2D-grid can only be simulated correctly when using the parallel strategy and that the other strategies lead to a break-down of the emergent pattern reported in the original paper. On the other hand, using the Heroes & Cowards game we showed that there exist models whose emergent patterns exhibit a stability under varying update-strategies. Intuitively we can say that this is due to the nature of the model specification which does not require specific orderings of actions but it would be interesting to put such intuitions on firm theoretical grounds in future research.

References

Bersini, H., & Detours, V. (1994). Asynchrony induces stability in cellular automata based models. In *In proceedings of artificial life IV* (pp. 382–387). Cambridge, MA: MIT Press.

Botta, N., Mandel, A., & Ionescu, C. (2010). *Time in discrete agent-based models of socio-economic systems* (Documents de travail du Cen-tre d'Economie de la Sorbonne No. 10076). Université Panthéon-Sorbonne (Paris 1), Centre d'Economie de la Sorbonne.

Clinger, W. D. (1981). *Foundations of actor semantics* (Tech. Rep.). Cambridge, MA: Massachusetts Institute of Technology.

Dawson, D., Siebers, P. O., & Vu, T. M. (2014, September). Opening Pandora's box: Some insight into the inner workings of an Agent-Based Simulation environment. In *2014 Federated conference on computer science and information systems* (pp. 1453–1460). https://doi.org/10.15439/2014F335.

Greif, I. (1975). *Semantics of communicating parallel processes* (Tech. Rep.). Cambridge, MA: Massachusetts Institute of Technology.

Hewitt, C., Bishop, P., & Steiger, R. (1973). A universal modular ACTOR formalism for artificial intelligence. In *Proceedings of the 3rd international joint conference on artificial intelligence* (pp. 235–245). San Francisco: Morgan Kaufmann Publishers Inc.

Huberman, B. A., & Glance, N. S. (1993, August). Evolutionary games and computer simulations. *Proceedings of the National Academy of Sciences, 90*(16), 7716–7718.

Jansen, M. (2012). *Introduction to agent-based modeling*. Retrieved from https://www.openabm.org/book/introduction-agent-based-modeling

Lysenko, M., D'souza, R., & Rahmani, K. (2008). *A framework for megascale agent based model simulations on the GPU*, Journal of Artificial Societies and Social Simulation (JASSS) 11(4):10. http://jasss.soc.surrey.ac.uk/11/4/10.html

Nowak, M. A., & May, R. M. (1992, October). Evolutionary games and spatial chaos. *Nature, 359*(6398), 826–829. https://doi.org/10.1038/359826a0.

Page, S. E. (1997, February). On incentives and up-dating in agent based models. *Computational Economics, 10*(1), 67–87. https://doi.org/10.1023/A:1008625524072.

Railsback, S., & Grimm, V. (2011). *Agent-based and individual-based modeling: A practical introduction*. Princeton, NJ/Oxford, UK: Princeton University Press.

Wilensky, U., & Rand, W. (2015). *An introduction to agent-based modeling: Modeling natural, social, and engineered complex systems with NET-logo*. Cambridge, MA/London, UK: MIT Press.

Wooldridge, M. (2009). *An introduction to multiagent systems* (2nd ed.). Hoboken, NJ: Wiley Publishing.

Yuxuan, J. (2016). *The agent-based simulation environment in Java* (Unpublished doctoral dissertation). University of Nottingham, School of Computer Science.

Chapter 4
Using Parallel Computing to Improve the Scalability of Models with BDI Agents

Patrick Taillandier, Mathieu Bourgais, Alexis Drogoul, and Laurent Vercouter

Introduction

The interest of using cognitive and emotional agents in social simulation has been showed by many works (Adam and Gaudou 2016; Truong et al. 2015) and more and more models tend to integrate such complex agents. This spreading is partly due to the integration in the classic agent-based modeling platforms such as Netlogo (Wilensky and Netlogo 1999) or GAMA (Gama website 2018; Grignard et al. 2013) of dedicated cognitive agent architectures that simplify the definition of cognitive agents through their modeling language. The architecture integrated in the GAMA platform is particularly rich: it does not only allow to define cognitive agents, but it links their cognition with an emotional module and a social-relation engine.

P. Taillandier (✉)
MIAT, Université de Toulouse, INRA, Castanet-Tolosan, France

UMR IDEES, CNRS, UNIROUEN, Rouen, France
e-mail: patrick.taillandier@inra.fr

M. Bourgais
UMR IDEES, CNRS, UNIROUEN, Rouen, France

Normandie Université, INSA, Rouen, France

UNIHAVRE, UNIROUEN, LITIS, Rouen, France
e-mail: mathieu.bourgais@insa-rouen.fr

A. Drogoul
UMI 209 UMMISCO, IRD/UPMC, Bondy, France

ICTLab, USTH, VAST, Hanoi, Vietnam
e-mail: alexis.drogoul@ird.fr

L. Vercouter
Normandie Université, INSA, Rouen, France

UNIHAVRE, UNIROUEN, LITIS, Rouen, France
e-mail: laurent.vercouter@insa-rouen.fr

© Springer Nature Switzerland AG 2019
D. Payne et al. (eds.), *Social Simulation for a Digital Society*, Springer
Proceedings in Complexity, https://doi.org/10.1007/978-3-030-30298-6_4

However, even with these architectures, an issue still open is the computing time they require. Indeed, as the studied systems are often composed of thousands of entities, the computation time of the proposed tools is a true concern. Moreover, most of existing and generic agent-based modeling platforms execute the simulation in a single thread/core (except for the visualization part), whereas recent computers integrate four cores or more, not to mention the possibility to use clusters and grids.

In order to address this problem and to benefit from multi-core architectures during the execution of simulations, we propose in this paper a new version of the BDI cognitive and emotional architecture integrated in the GAMA platform (Bourgais et al. 2016; Caillou et al. 2015; Taillandier et al. 2016), which allows to parallelize in a transparent way most of the required computations.

This article is structured as follows: First, we present the state of the art on the existing cognitive architectures integrated in simulation platforms as well as the different works that concern the parallelization of agents in simulations. Next, we present the parallel version of the cognitive architecture. Then, we present experiments carried out with this architecture before we finish with a conclusion.

Related Works

Cognitive Agent Architectures

These last years, many cognitive architectures dedicated to agent-based simulations have been proposed. A lot of them are based on the BDI paradigm (Bratman 1987). This paradigm proposes a straightforward formalization of the human reasoning through intuitive concepts (beliefs, desires and intentions).

In order to ease the development of BDI agents, some works have proposed to integrate such an architecture into a specific framework. The most famous are the Procedural Reasoning System (PRS) (Myers 1997), JACK (Howden et al. 2001) and JADEX (Pokahr et al. 2005). Unfortunately, these frameworks still require a high level in computer science and Artificial Intelligence to be used.

To face this difficulty, other authors have proposed to directly integrate a BDI architecture into generic agent-based modeling platforms that are often used for social simulations. For example, an extension of NetLogo proposes a simplified BDI architecture for educational purposes (Sakellariou et al. 2008). Agents have beliefs, intentions and ways to answer to intentions. Another integration of the BDI architecture concerns the GAMA platform. The developed architectures have been created to be as complete as possible and available to a non-expert public. It gives agents a beliefs base, a desires base, an intentions base and a reasoning engine as well as an emotional engine. Some works have already shown that the use of this architecture eases the modeling of human beings (Adam et al. 2017; Taillandier et al. 2016). However, if some previous works such as (Truong et al. 2015) showed that the architecture enables to simulate the behaviors of thousands of agents, its limitation of execution in a unique thread (and thus one core) raises a scalability question. To face this difficulty, we propose a new version of this architecture that allows to parallelize some of its computations on different computer cores.

Parallelization of a Simulation

If defining an experiment that runs a set of simulations on several computer cores (one simulation per core) is a common practice, in particular with dedicated frameworks such as OpenMOLE (Rey-Coyrehourcq et al. 2013), the parallelization of a unique simulation is much more uncommon. A reason is that in agent-based simulations, the agents often interact with many others, which makes the decomposition of a simulation difficult. However, some works propose solutions to tackle this difficulty.

Among these works, some proposes ad hoc workarounds dedicated to a specific model or to a specific application domain such as (Marilleau et al. 2012; Parker 2007) and thus cannot be applied for all types of applications. Other works propose sets of generic algorithms or toolkits such as MCMAS (Laville et al. 2013) that integrate classic algorithms (diffusion, path-finding...) for agent-based simulation. Although this type of toolkits cannot be directly used by modelers that are not computer scientists, they can be used as basic components to improve modeling platforms. At last, some works propose dedicated environments in which modelers have to respect some specific constraints to define their agents (Collier and North 2011; Marilleau et al. 2012; Richmond et al. 2010). While these frameworks allow to obtain high scalability results, their use is far beyond the reach of most of modelers, in particular when modelers have to define cognitive agents.

To conclude, although solutions already exist to parallelize a simulation, their lack of genericity and their difficulty of use explain why only few modelers use them. Our paper proposes a way to address this issue by offering the possibility to modelers to parallelize in a transparent way some of the computations required by the BDI architecture of GAMA.

Parallel Cognitive and Emotional Architecture

Our architecture is based on the cognitive architecture integrated in GAMA (*simple_bdi*) (Bourgais et al. 2016; Caillou et al. 2015; Taillandier et al. 2016), which aims at letting modelers use the BDI paradigm to define their cognitive agents. The main strength of this architecture is to be very versatile (a minimal core with a lot of optional features) and usable even by a non-expert public (Adam et al. 2017; Taillandier et al. 2016) through the use of the GAML language (the modeling language of the GAMA platform). An important objective of this work was to propose a new version of the architecture allowing to parallelize the computation as easy to use as the current one with the minimum of new keywords. Thus, we defined the new architecture such that using it instead of the classic one just requires the modeler to change the name of the control architecture from *simple_bdi* to *parallel_bdi*.

Like the classic simple BDI architecture, our architecture provides agents with six bases:

- **Belief** (what it thinks): the internal knowledge the agent has about the world.
- **Desire** (what it wants): the objectives that the agent would like to accomplish.
- **Intentions** (what it is doing): the desire that the agent has currently chosen to fulfill.
- **Uncertainty** (what it expects to be true): the uncertain events that the agent expects.
- **Emotions** (what it feels): the emotions of the agent. In this work, emotions are based on the OCC theory (Ortony et al. 1990) which means they are seen as a valued answer to the cognitive appraisal of a situation.
- **Social relations** (its relationship with others): the social links the agent has with other agents. Here, social links are defined thanks to the dimensional model of interpersonal relation of Svennevig (2000). A social link contains information about liking, dominance, solidarity and familiarity towards another agent.

The reasoning engine of the agents with the architecture is also almost similar to the one of simple BDI, except that it integrates the notion of parallelization. Indeed, in GAMA, the agents are scheduled as follows: at each simulation step, all agents are activated one by one according to a given order, which is by default their order of creation, but that can be simply modified. Once a BDI agent is activated, it executes its complete cycle of reasoning (perceives the world, chooses and executes plans…). One of the major modifications of our new architecture is to split this cycle in sub-steps that can be either parallelized (i.e. distribution of the computation of this sub-step on the different cores of the computer) or not. Thus, each sub-step can be executed sequentially (while keeping the activation order defined by the scheduler) or in parallel (all BDI agents execute the sub-step at the same time).

This reasoning engine is composed of eight sub-steps (Fig. 4.1).

Step 1. Degradation of predicates: The first step of the reasoning cycle is the degradation of the knowledge of the agent. The predicates stored in the cognitive bases are reduced in lifetime. This step is automatically parallelized.

Step 2. Degradation of emotions: The second step of the reasoning cycle is the degradation of the emotions of the agent. The intensity of agent's emotions is reduced by their decay value. Like Step 1, this step is automatically parallelized.

Step 3. Perception: The third step of the reasoning cycle of the agent is the perception of the environment. The perceptions are described by the modeler through *perceive* blocks. Each perceive block is applied and can participate to update the beliefs of the agent and create social links with other agents. Example of definition of perceive blocks through the GAML language can be found in

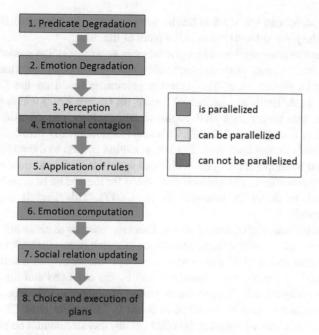

Fig. 4.1 Schema of our reasoning engine

Taillandier et al. (2016). As the modeler has the possibility to define interactions with other agents in a perceive block that can introduce a parallel access to a variable (with for instance a simultaneous writing and reading of the variable), we let the modeler choose if he/she wants to parallelize each block or not. By default, all the perceive blocks are parallelized, but a modeler can specify that he/she wants a perceive to be not parallelized by just setting to *false* the *parallel* facet of the perceive statement.

Step 4. Emotional contagion: The emotional contagion processes are defined inside a perceive block and allows an agent to update its emotions according to the emotions of the nearby agents. As this step requires each agent to simultaneously modify its emotion bases and check the emotion bases of the nearby agents, this step cannot be parallelized. As a consequence, the *parallel* facet of a perceive block with emotion contagion statements inside should always be set to *false*.

Step 5. Inference rules: In this step, the agent applies its inference rules to manage the belief, desire and emotion bases according to its previous beliefs and desires (and eventually to other events). This step gives a dynamic to the overall behavior as the agent can act according to a change in the environment. These inference rules can also be influenced by emotions or social relations. Example of definition of rules through the GAML language can be found in Taillandier et al. (2016). Like for the perceive block, the modeler has the possibility in a rule to define interactions with other agents. So, we let the modeler choose if he/she wants to parallelize each rule or not. By default, all the rules are parallelized, but a modeler can specify that he/she wants a rule to be not parallelized by just setting the *false* value to the *parallel* facet of the rule.

Step 6. Emotion computation: The optional emotion computation module allows to automatically create, with no intervention from the modeler, emotions based on the agent's mental state. This creation process is based on the OCC theory (Ortony et al. 1990) and its logical formalism (Adam 2007) which proposes to integrate this theory in a BDI cognitive architecture. The creation process is detailed in Bourgais et al. (2017). This step is automatically parallelized.

Step 7. Social relation updating: The social engine is used to dynamically update the social relations of the agent with other agents according to the emotions and the cognition. The computation formulations for the update of social relations is described in detail in Bourgais et al. (2017). This step is automatically parallelized.

Step 8. Choice and application of plans: This step consists in choosing one or several plans and to execute them. The choice of a plan passes through the choice of an intention and then of a plan to achieve this intention. The whole process is influenced by the cognitive bases but also by the emotions and the social relations of the agent and, through the execution of plans, can influence these bases. This cognitive engine is described in detail in Taillandier et al. (2016). As the execution of plans will impact the other agents, this step cannot be parallelized.

This architecture (with its new extension) is already available within the last version (1.7) of the GAMA platform (Grignard et al. 2013). As stated in the previous paragraphs, modelers can easily use it – with just few lines of codes (by choosing the *parallel_bdi* control architecture and optionally setting the parallel of some perceive and rule statements to false) – through the GAMA dedicated modeling language.

Case Study

In order to validate the parallel BDI architecture, we test its results on the three models provided with the BDI architecture:

- **Gold Miners**: this model concerns gold miners that try to find and extract gold nuggets (see Fig. 4.2). A Miner agent wanders around to find gold nuggets. When it perceives some gold nuggets, it stores this information and begins to extract the closest one then, it brings back the gold nugget to the base and goes to search for another gold nugget.
- **Firefighters**: this model concerns the actions of firefighters against fires (see Fig. 4.3). In this model, a firefighter agent patrols, looking for fires. When it finds one, it tries to extinguish it by dropping water, and when it has no more water, it goes to the nearest lake to refill its water tank.
- **City Escape**: this model concerns the evacuation of drivers from a city after a hazard (see Fig. 4.4). In the studied city, a factory of chemical products is on fire and some products are spread into the air. If a driver smells the toxic gas, he/she

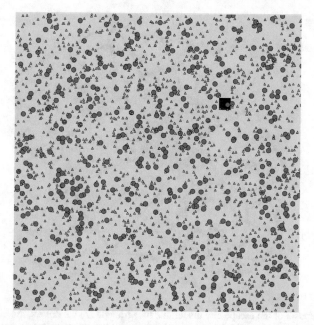

Fig. 4.2 Snapshot of the gold miner model: the green circles are the miners, the yellow triangles the gold nuggets, and the black square is the base

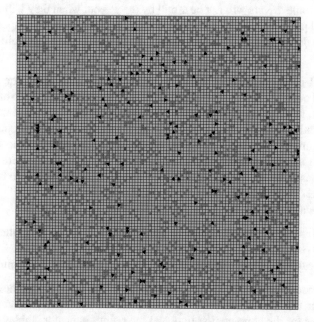

Fig. 4.3 Snapshot of the firefighter model: the black triangles are the firefighters, the red circle the fires, and the blue squares the lakes

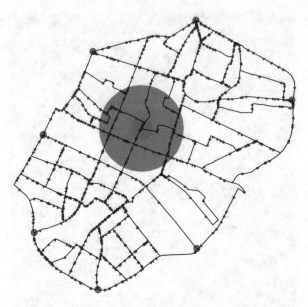

Fig. 4.4 Snapshot of the escape city model: the blue triangle are the drivers, the magenta circles the shelters, the green circle the fire perception radius, and the red circle the gas perception radius

will think that it is possible that a catastrophe happened, so he/she will go to a shelter. Some drivers will not be afraid by the gas but when they will see the fire, they will panic and go to the shelters twice faster than the normal speed. In addition, if a driver sees other drivers fleeing, he/she will have a probability to flee to a shelter.

Table 4.1 gives quantitative information about the three models. The experiment was carried out on a simple i7 Windows laptop with 4 cores and 32 Gb of RAM. For each model, we stopped the simulation after 200 simulation steps and we measured the computation time (without the display part – the models are executed in batch mode) and an indicator to illustrate the output difference between the simple BDI architecture and the parallel one. Indeed, as the parallel architecture introduces modifications about the general scheduler of GAMA, it is important to have an idea of the impact of this modification on the simulation output.

The indicator was defined as follow for the three models:

- **Gold Miners**: number of remaining gold nuggets after 200 simulation steps
- **Firefighters**: number of remaining fires after 200 simulation steps
- **City Escape**: number of drivers that have not escaped after 200 simulation steps

As the models are stochastic, we ran each model ten times (using the same series of seeds for the *simple_bdi* and *parallel_bdi* architectures) and we computed the mean values for the computation time and the output indicator. Tables 4.2 and 4.3 present respectively the results obtained in terms of output indicators and computa-

Table 4.1 Quantitative information concerning the three models

Data	Gold miner	Fire fighter	Escape city
Number of cognitive agents	500	200	1000
Number of plans per agents	4	2	3
Number of perceive per agents	1	2	3
Number of rules per agents	2	2	2
Use of emotions	False	False	True

Table 4.2 Output indicator with the three models (mean results for ten simulations)

Model	Using simple_bdi	Using parallel_bdi
Gold miner	1299.6	1303.3
Fire fighter	782.7	783.2
Escape city	964	978.2

Table 4.3 Computation times (in ms) with the three models (mean results for ten simulations)

Model	Using simple_bdi	Using parallel_bdi
Gold miner	8746 ms	5394 ms
Fire fighter	10,455 ms	5862 ms
Escape city	6084 ms	5033 ms

tion time for the three models with the two architectures. The first result that we can observe is that the difference for the three models in terms of output indicators is very low, which means that the use of our architecture had a insignificant impact on the simulation results. In terms of computation time, using the parallel BDI architecture allowed for the two gold miner model to decrease the computation time by 38%, the fire fighter model by 43% and the escape city model by 17%.

The fact that the improvement of the computation times of the escape city model was less important whereas it integrates some features linked to the emotional modules that are parallelized can be explained by the fact that the plans carried out by the agents are more time-consuming. Indeed, the agents have first to compute the shortest path between their current location and their target, and then to move along the polyline roads. It could have been possible to modify a bit the model to integrate the shortest path computation inside the perception rather than in a plan to get a better result, but as the objective of the experiment was to show that the use the parallel architecture allows to improve the results by just changing the control architecture (without modifications of the model), we did not. Nevertheless, the results obtained are promising considering the fact that the experiment was carried out on a simple laptop computer. The improvement of the computation time will be far more important on a more powerful computer (with more cores).

Conclusion

In this paper, we have presented a parallelized version of the cognitive architecture integrated in the GAMA platform. As shown by the experiments carried out on three models, the use of the parallel BDI architecture that just requires to change one or two words in the code allows to significantly decrease the computation time required even on a simple laptop computer without altering the simulation results.

Acknowledgments This work is part of the ACTEUR ("Spatial Cognitive Agents for Urban Dynamics and Risk Studies") research project funded by the French National Research Agency.

References

Adam, C. (2007). *Emotions: From psychological theories to logical formalization and implementation in a BDI agent.*

Adam, C., & Gaudou, B. (2016). BDI agents in social simulations: A survey. *The Knowledge Engineering Review, 31*(03), 207–238.

Adam, C., Taillandier, P., & Dugdale, J. (2017). Comparing agent architectures in social simulation: BDI agents versus finite-state machines. In *Hawaii International Conference on System Sciences (HICSS).*

Bourgais, M., Taillandier, P., & Vercouter, L. (2016). An agent architecture coupling cognition and emotions for simulation of complex systems. In *Social simulation conference.*

Bourgais, M., Taillandier, P., & Vercouter, L. (2017). Enhancing the behavior of agents in social simulations with emotions and social relations. In *MABS* (to be published).

Bratman, M. (1987). *Intentions, plans, and practical reason.* Cambridge, MA: Harvard University Press.

Caillou, P., Gaudou, B., Grignard, A., Truong, C. Q., & Taillandier, P. (2015). A simple-to-use bdi architecture for agent-based modeling and simulation. In *ESSA.*

Collier, N., & North, M. (2011). Repast HPC: A platform for large-scale agent-based modeling. In *Large-scale computing techniques for complex system simulations* (pp. 81–110).

Gama website. (2018). https://gama-platform.github.io/

Grignard, A., Taillandier, P., Gaudou, B., Vo, D. A., Huynh, N. Q., & Drogoul, A. (2013). Gama 1.6: Advancing the art of complex agent-based modeling and simulation. In *International conference on principles and practice of multi-agent systems.* Berlin, Germany: Springer.

Howden, N., Rönnquist, R., Hodgson, A., & Lucas, A. (2001). Jack intelligent agents-summary of an agent infrastructure. In *5th International conference on autonomous agents.*

Laville, G., Mazouzi, K., Lang, C., Marilleau, N., Herrmann, B., & Philippe, L. (2013). MCMAS: A toolkit to benefit from many-core architecture in agent-based simulation. In *European conference on parallel processing* (pp. 544–554). Berlin Heidelberg: Springer.

Marilleau, N., Lang, C., Chatonnay, P., & Philippe, L. (2012). RAFALE-SP: A methodology to design and simulate geographical mobility. *Studia Informatica Universalis, 10*(1), 38–76.

Myers, K. L. (1997). *User guide for the procedural reasoning system.* SRI International AI Center Technical Report.

Ortony, A., Clore, G. L., & Collins, A. (1990). *The cognitive structure of emotions.* Cambridge, UK: Cambridge University Press.

Parker, J. (2007). A flexible, large-scale, distributed agent based epidemic model. In Simulation conference, 2007 Winter, pages 1543–1547. Piscataway, NJ: IEEE, 2007.

Pokahr, A., Braubach, L., & Lamersdorf, W. (2005). Jadex: A BDI reasoning engine. In R. H. Bordini, M. Dastani, J. Dix, & A. E. Fallah-Seghrouchni (Eds.), *Multi-agent programming: Languages, platforms and applications* (pp. 149–174). Boston, MA: Springer.

Rey-Coyrehourcq, S., Reuillon, R., & Leclaire, M. (2013). Openmole, a workflow engine specifically tailored for the distributed exploration of simulation models. *Future Generation Computer Systems, 29*(8), 1981–1990.

Richmond, P., Walker, D., Coakley, S., & Romano, D. (2010). High performance cellular level agent-based simulation with flame for the GPU. *Briefings in Bioinformatics, 11*(3), 334–347.

Sakellariou, I., Kefalas, P., & Stamatopoulou, I. (2008). Enhancing netlogo to simulate BDI communicating agents. In J. Darzentas, G. A. Vouros, S. Vosinakis, & A. Arnellos (Eds.), *Artificial intelligence: Theories, models and applications* (pp. 263–275). Berlin, Heidelberg: Springer.

Svennevig, J. (2000). *Getting acquainted in conversation: A study of initial interactions.* Philadelphia: John Benjamins Publishing.

Taillandier, P., Bourgais, M., Caillou, P., Adam, C., & Gaudou, B. (2016). A BDI agent architecture for the GAMA modeling and simulation platform. In *MABS 2016 multi-agent-based simulation.*

Truong, Q. C., Taillandier, P., Gaudou, B., Vo, M. Q., Nguyen, T. H., & Drogoul, A. (2015). Exploring agent architectures for farmer behavior in land-use change. A case study in coastal area of the Vietnamese Mekong delta. In *International workshop on multi-agent systems and agent-based simulation* (pp. 146–158). Cham, Switzerland: Springer.

Wilensky, U., & Netlogo, I. E. (1999). *Center for connected learning and computer-based modeling.* Evanston, IL: Northwestern University.

Part II
Applications to Norm Diffusion and Collective Action

Chapter 5
Information Diffusion as a Mechanism for Natural Evolution of Social Networks

Kyle Bahr and Masami Nakagawa

Introduction

Recent work in the area of social license to operate (SLO)—and earlier work in the more general field of corporate social responsibility (CSR)—represents an attempt to understand the individual and group behaviors that are relevant to the development of local, resource-based, economic projects (Parsons and Moffat 2014; Prno and Slocombe 2012; Thomson and Boutilier 2011). This is an area with high potential for elucidation via agent-based modeling, and some work has already been done in that direction (Bahr and Nakagawa 2017; Nakagawa et al. 2012; Okada 2011). Most often, these projects are in the mining and oil and gas sectors, which are large, highly visible, resource intensive industries with a high potential for environmental, social, and economic impacts to the local areas in which they operate. Stakeholder analysis is one of the most useful tools available to both managers and communities who are involved in these kinds of projects and interested in understanding individual and group behaviors. The focus of stakeholder analysis on identifying individual actors and quantifying their opinions and influence is a key feature, which serves to clarify some of the more fluid elements of what is traditionally known as "the community" in CSR literature (Freeman and Mc Vea 1984).

One major problem with "the community" as a construct for formal analysis is that it has ambiguous and dynamic boundaries and inclusion/exclusion from the group is based on varying criteria which may or may not be relevant to a given stakeholder. The benefit of using stakeholder analysis, therefore, is the clear definition of the group of interest as those who may affect or be affected by the actions of

K. Bahr (✉)
Tohoku University, Sendai, Japan
e-mail: kyle@geo.kankyo.tohoku.ac.jp

M. Nakagawa
Colorado School of Mines, Golden, CO, USA
e-mail: mnakagaw@mines.edu

© Springer Nature Switzerland AG 2019
D. Payne et al. (eds.), *Social Simulation for a Digital Society*, Springer
Proceedings in Complexity, https://doi.org/10.1007/978-3-030-30298-6_5

a company or industry (Mitchell et al. 1997). This may include both geographically local and external actors, who often play a crucial role in the granting or withholding of the social license. The clear boundaries of who qualifies as stakeholders also allows for the use of network analysis, which can aid in understanding the relationships between stakeholders and determining the relative importance of individuals within the network.

With stakeholder analysis as a foundational basis, Bahr and Nakagawa have proposed an agent-based model for understanding the diffusion of opinion within a stakeholder group (Bahr and Nakagawa 2017; Nakagawa et al. 2012). Agent-based modeling is a useful method in which computational analogues (agents) are given characteristics and interaction rules on an individual scale, and emergent phenomena are observed on the aggregate level (see, for example, Gilbert (2008)). Agent-based diffusion models have great practical applicability for studying the generation and evolution of stakeholder networks. Previous social diffusion models include those of Axelrod (1997), Deffuant et al. (2000), and Hegselmann and Krause (2002). Although there is much literature available on algorithms for the generation and analysis of social networks, these approaches focus only on generation techniques and not on the evolution and emergence of network structures themselves (Albert and Barabási 2002; Barabási and Albert 1999; Erdös and Rényi 1959; Watts and Strogatz 1998). In addition, little is shown in the literature with regard to the decay of network links (Preusse et al. 2014; Roberts and Dunbar 2015).

In this paper, we will demonstrate how network evolution logic can be used to better capture the role of interpersonal influence in opinion diffusion, and how such a model may then be used to generate real-world stakeholder networks and analyze the stability and relevance of those networks in social license applications.

Model Design

The original model proposed by Nakagawa and Bahr in 2012 and the subsequent enhancements made to the model in 2016 serve as the basic framework upon which the model presented here is built. As such, it is appropriate to give a simple, general description of the original models before describing the network generation logic that is the subject of this current research. For a more detailed treatment on the workings of the original models (see Bahr and Nakagawa 2017; Nakagawa et al. 2012).

The Bahr and Nakagawa models use agent profile characteristics including "consent" as an individual state variable, "opinion" as a transmission commodity, and "influence" as measure of each agent's ability to transfer their opinion to other agents. Consent as a variable is a dynamic measure of individual judgment on the legitimacy of a focal organization (the company, industry, or project seeking the SLO). It is discretized on a 5-point integer scale ranging from -2 (strong opposition to the focal organization) to $+2$ (strong support of the focal organization) and it is a state variable, dependent on the value of the opinion variable.

Opinion in the model is an analogue for the information that each stakeholder uses to determine their own level of consent. It is a vector quantity with both magnitude (the depth of feeling) and direction (support or opposition). As agents internalize positive opinion toward the focal organization, their opinion of its legitimacy increases and the consent variable changes in the positive direction at some threshold of opinion. The opposite is true for negative opinion toward the focal organization. For modeling convenience, consent categories are tied to a standard of 10 opinion units, although in real-world situations this may be arbitrary. In other words, agents having opinion values ranging from 0 to 10 fall in the −2 (high opposition) consent category, values from 11 to 20 denote a −1 (low opposition) category, continuing up to opinion levels from 41 to 50 representing a +2 (high support) consent category.

Influence is a description of the weight that any given agent gives to the opinion of any other agent that they may come in contact with. Influence in the original models is a static variable assigned to agents based on some user-determined distribution. Like consent, it is also discretized into 5 categories ranging numerically from −2 to +2 (low to high influence). For the purpose of understanding the real-world meanings of these categories, each influence category has its own explanatory description: "marginal" (−2), "vulnerable" (−1), "influential" (0), "strong" (+1), and "dominant" (+2). The weight given to an agent's opinion is related to the magnitude (absolute value) of the differential in influence between two interacting agents, with 0.25 increments between each adjacent influence category. For example, when a marginal agent meets with a strong agent, the weight of the strong agent's opinion is 0.75 (e.g. |+1 − (−2)| × 0.25).

The central decision-making function of the model is given by Eq. 5.1, in which two agents, i and j, having influence I_i and I_j and consent levels C_i and C_j respectively, meet and interact with the potential for opinion gain or loss Δx_i and Δx_j out of a total potential given amount X:

$$\text{For } I_i > I_j \begin{cases} \Delta x_j = 0.25 \left| I_i - I_j \right| \text{sgn}\left(C_i - C_j \right) X \\ \Delta x_i = N \Delta x_j \end{cases} \tag{5.1}$$

Note that in the lower half of Eq. 5.1 the change in opinion of agent i is merely the change in opinion of agent j multiplied by a factor, N. N is the addition of Bahr and Nakagawa's 2017 paper which, broadly speaking, allowed for opinion to be exchanged as a "dialogue" in proportion to the relative influence of two interacting agents. N is a static, contextually dependent and globally defined parameter ranging from 0 to 1 in 0.01 increments.

Using the above description of model attributes and variables, the model results are characterized in one of two ways; either a consensus is achieved among the agents (i.e. all agents reach the same consent category), or an equilibrium state is reached in which the proportion of agents in each consent category does not change.

In this paper, we introduce five pieces of modeling logic to Bahr and Nakagawa's opinion diffusion model that allows for the creation of a plausible agent-based analogue for real-world dynamic network generation: Formation of network links through agent interaction; centrality to replace static influence; preference for attachment; equivalent influence logic; and link decay.

Link Formation

As each agent meets and interacts with other agents, it makes decisions about how its opinion and consent variables change based on the characteristics of the agents with which it interacts. As opinion is passed between two agents that are interacting for the first time, a directional link is created from source node to sink node (high influence agent to low influence agent), and the newly-created link has a strength that is proportional to the amount of opinion transferred. If N from Eq. 5.1 has a non-zero value, then a link will also be created in the opposite direction (j to i) with strength proportional to Δx_i. If two agents with an existing link interact, that link is strengthened by the amount of opinion transferred in that interaction. Figure 5.1 shows the link formation procedure.

Centrality as Influence

Network centrality can be used as a measure of the importance that each node has in the overall system, which makes it a natural choice for determining each agent's influence level at a given timestep. At each timestep, the network model measures the centrality score of each agent and sets its influence variable accordingly. This is then used in determining the amount of opinion that is transferred between two

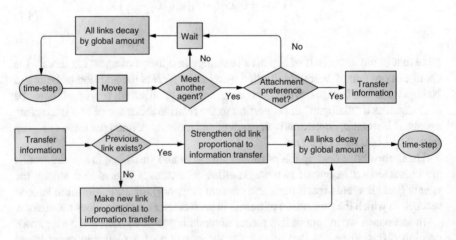

Fig. 5.1 Link formation protocol

interacting agents within that timestep. Network centrality can be thought of and measured in many different ways, depending on the assumptions one makes about how traffic flows from one node to another through the network. Borgatti has outlined a useful framework for determining what measures best suit a network of interest based on trajectories (path, walk, trail, or geodesic distance) and transfer mechanisms (transfer or replication, series or parallel) of traffic (Borgatti 2005). In the case of the opinion diffusion model, the most appropriate set of assumptions is that information is transferred via the serial walk, as opinion is a divisible package that can revisit nodes and links, and agents are more likely to meet one-on-one than as a group in a given timestep. Given those assumptions, it was determined that eigenvector centrality and Freeman degree centrality are the most[1] appropriate candidates for measuring network centrality in this case.

Eigenvector centrality has the advantage that it reflects the global importance of a node in the overall network, but it is computationally expensive for large matrices and it requires a minimum density in the adjacency matrix—a criterion that is not always met in observed model states. In contrast, while Freeman degree centrality is fairly primitive in its ability to capture centrality behavior, it is correspondingly less complex to calculate and does not require any minimum density in the adjacency matrix. Additionally, because Freeman degree is a measure of local influence, it requires only the assumption that agents are aware of the connections of themselves and their potential interaction partners. This is a realistic assumption that is validated by the work of Kearns and others, who have found that knowledge of the overall structure of the network is often not available to individual nodes, while some knowledge of the ego networks of nearest neighbors often is (Judd and Kearns 2008; Kearns 2012). These considerations ultimately led to the decision to use Freeman degree centrality for updating the influence variables of each agent at each timestep.

Because the influence variables of the agents were defined as numerical integer values from -2 to 2, describing the five influence categories (dominant, strong, influential, vulnerable, and marginal) in the original model, it was necessary to normalize the numerical centrality scores relative to some maximum centrality (μ_{max}), in order to discretize those scores into the five influence categories. This was done using a linear relationship in 20% increments of μ_{max} ($\mu_i > 0.2\mu_{max} \rightarrow$ marginal, etc.). The reference point (μ_{max}) was chosen by assuming that the most well-connected node is the most dominant of the network, and setting each other node's influence level relative to that of the most dominant. The most well connected node has the most social capital of all the nodes, which makes it the most influential, having the most resources to mobilize in favor of spreading its opinion. The assumption here is that each node does not need to know the overall size of the network, only who is the most well connected node. In practice, this reference point may skew the results

---

[1] As Borgatti points out, most existing centrality measures are best suited for parallel duplication and geodesic distance assumptions. However, both Freeman degree and eigenvector centrality have characteristics that make them reasonable measures for the purposes of this model.

somewhat, as the first agent to have an interaction will immediately be dominant while all other agents will continue to be marginal. This is easily remedied by using initial conditions in which a seed network of some description already exists.

Preferential Attachment

In Bahr and Nakagawa's earlier models, there is no mechanism for agents to manifest a preference for interaction with other agents; any time two agents meet, they have an interaction. In reality, individuals do not simply react with every person that they meet. In creating networks dynamically through agent interaction and information exchange, networks in which all agents have an equal probability of interaction with all other agents tend to produce very dense networks with no core-periphery structure. In contrast, many real-world stakeholder networks exhibit a high degree of core-periphery behavior, in which stakeholders are either part of a highly connected central group that bridges and bonds other stakeholders, or are part of a less-connected set of individuals that are connected to other stakeholders through the more central actors (Boutilier 2009).

In order to reflect the real-world tendency of stakeholders to selectively create links with more popular (well-connected) stakeholders, a preferential attachment algorithm was implemented. The procedure is initiated at the beginning of an interaction if $AP \neq 0$, where AP is the globally-set attachment preference variable. First, the agent with a higher degree centrality (i) compares their preference for attachment with the lower influence agent's (j) centrality normalized by μ_{max}. If $\mu_j/\mu_{max} > AP$, then agent i initiates an interaction. If not, then the lower influence agent, j, generates a random float between 0 and 1. If that value is higher than AP, then agent j initiates an interaction and agent i accepts, even though agent j's centrality is too low. The probability of an interaction in the case where $\mu_j/\mu_{max} \leq AP$ is inversely proportional to how large the attachment preference is (high preference = low probability of interaction). For "peers" (see the next section) who do not already have links with each other, one agent goes through the above procedure, and if no interaction occurs then the other agent goes through the procedure. This model logic allows agents to constrain the relationships they build and the opinion they give based on the perceived existence of some benefit for creating a relationship.

Equivalent Influence

The mechanism for opinion exchange in the original model is dependent on the influence differential between two agents (see Eq. 5.1). This reflects a real-world assumption that individual actors will not be persuaded by others unless there is some direct benefit (or disadvantage) that will be the consequence of doing so (or not doing so). This assumption is limiting, in two ways. As a real-world analogy, it fails to reflect the possibility of persuasion through reasoned discussion with peers, instead suggesting either opportunism or intimidation as the only motivations for opinion change. As a modeling tool it negates the possibility of an interaction between two agents within the same influence category, despite the very high probability of a meeting between two such agents. This has the effect of biasing the model results toward a state of neutral consensus.

A more nuanced assumption would be that individuals, while still persuading and being persuaded on the grounds of relative influence, also change their opinions based on the other factors and characteristics of their peers, including, but not limited to, appeals to pathos, trust between friends, the logic of the argument, etc. A teacher, for example, has a certain level of credibility by virtue of the imbalance in position between teacher and student, but two equally influential students may still convince each other of opposing positions through the use of rhetorical skill. In the latter case, a kind of temporally localized influence differential is created, in which one student changes the mind of the other to at least some degree based on the advantages gained by what passes between them at the time.

In order to accommodate this more nuanced description of opinion dynamics, modeling logic that allows for the transfer of opinion between two agents based on some temporally localized advantage was implemented as an addition to the original model. This logic is only called during the specific circumstance of influence-equivalent agents meeting in the interaction space and is limited to a single timestep. When such an interaction occurs, each of the interacting agents is assigned a number from a normal distribution, and the flow of opinion for that interaction is proportional to this new differential according to the same rules as Eq. 5.1, wherein global influence I_i and I_j are replaced with this new local influence i_i and i_j. It is important to note that if those same agents meet again while occupying the same global influence category, the equivalent influence logic is implemented again so that sometimes opinion flows from i to j, and vice versa.

Because the localized influences are drawn from a normal distribution, the amount of opinion exchanged is usually small, representing a more gradual shift in opinion that nonetheless allows agents to form network links with each other (the strength of the link still being proportional to the amount of opinion exchanged between agents). The ability to form network links with each other also contributes to the ability of each agent to ascend to the next global influence category, depending on the number of links each already has.

In order to test the effects of this new logic, a set of 100 model runs was performed with initial conditions of 100 agents uniformly distributed over both the consent and influence categories. This resulted in each of the five consent equilibrium states (high blocker consensus, blocker consensus, etc.) being reached a certain proportion of the simulation runs. Figure 5.2 shows a histogram of the results both *with* (light gray bars) and *without* (dark gray bars) the implementation of the equivalent influence logic. In both cases the agents are distributed normally over the consent categories as expected from the initial conditions, but the equivalent influence implementation has the effect of shortening the peak and spreading the distribution more widely. This means that the temporal localization of influence for

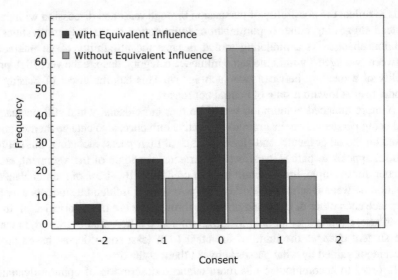

Fig. 5.2 Model outcomes with and without equivalent influence logic

equivalent influence agents acts in opposition to the previously mentioned neutrality bias. This is expected and even desirable from the modeling standpoint, because it provides a mechanism for dampening the effects of bidirectionality introduced in earlier models. The working interpretation of this result is that the bottom-up effect of peer-to-peer interactions is to produce a more broad spectrum in social license probabilities than the top-down effect of norms strictly enforced by a vertical influence structure.

Link Decay

The concept of link decay was introduced in order to balance the creation of new network links in the model. Essentially, it is a mechanism whereby the rate at which links disappear from the network can be controlled. It was implemented in the model by subtracting a globally set amount from the strength of each link during each timestep. This addition simulates the idea that relationships will deteriorate if they are not maintained, and how quickly they deteriorate is proportional to the strength of the relationship. For example, a one-time collaboration between researchers is likely to result in a relationship that disappears much more quickly than a formal agreement between two research institutions, especially if it is not renewed through future projects. If a one-time collaboration is renewed, on the other hand, the relationship may eventually grow to the level where it may be reinforced through such an institution-level agreement. The rate at which links decay is hypothesized to be contextually dependent, and must therefore be calibrated for a given network. Additionally, the overall rate of network development plays a role in the structure of the final (consensus or equilibrium) network. Without decay, links are created at a rate which increases monotonically until all agents are connected or a consensus is reached (Fig. 5.3a). With decay, the number of links will reach a stable equilibrium when the rate of formation of new links is matched by the rate of decay of old links (Fig. 5.3b).

Model Proof of Concept: Social License to Operate

After the implementation of the link formation, centrality, preferential attachment, equivalent influence and link decay logic as outlined above, the network model was subjected to a parametric study over the link decay and preference for attachment

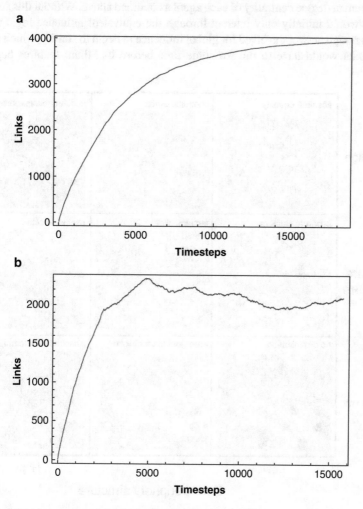

Fig. 5.3 (a) Link formation without link decay; (b) Link formation with link decay

variables in order to test its internal consistency and ability to recreate an observed network structure. Boutilier and Thomson (2009) have proposed a framework for categorizing observed stakeholder networks based on the characteristics of closure (density) and core-periphery structure (see Fig. 5.4). It is theorized that the features of closure and core-periphery structure determine the durability or resilience of the social license; closure, by determining the extent to which stakeholders are connected to each other and therefore able to organize and communicate, and core-periphery structure by determining the extent to which the opinion of the more dominant stakeholders represent the opinion of the less dominant stakeholders.

The model was initialized with the "factions" network—the central network from Fig. 5.4—as a seed structure for the model to build from. This was done so that there would be some initial distribution in the influence variable, which is based on the Freeman degree centrality of each agent as outlined above. Without this seed, all agents would initially only interact through the equivalent influence logic until a dense enough network evolved for global influence to begin to stratify, which means the model would have to run for some time before its salient features began to

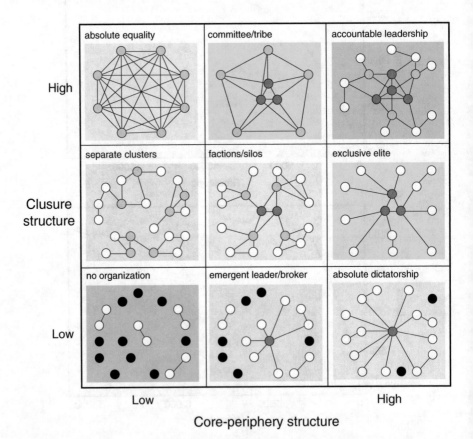

Fig. 5.4 Stakeholder network templates from Boutilier (Boutilier and Thomson 2009)

manifest. Additionally, networks of "no organization" like that shown in the bottom left of Fig. 5.4 are rare and transitory in practice, and there is usually already an existing network structure of some kind at the point where social license is measured.

The decay rate was calibrated to a scale that allowed for it to affect the network without annihilating the structure altogether; if the rate is too low, links would never disappear, and if it is too high, links would disappear before an overall structure could emerge. This calibration was performed by carefully balancing the rate of formation of links with the rate of decay of links, using the "perfect equality" and "no organization" networks (upper and lower left structures of Fig. 5.4, respectively) as upper and lower boundaries for the model.

Once the model was calibrated, a series of experiments were performed in which the model was run with a combination of link decay and attachment preference values. The model was run until either a consensus or equilibrium was reached. Each parametric combination was run 100 times with a different random generator seed in order to explore the probability of a given outcome.

Model Validation and Discussion

In order for the agent-based network model to be effective in forecasting the durability of a social license, it must be able to reproduce the structures shown in Fig. 5.5. Figure 5.6 shows the comparable network structures produced by the agent-based network model. Each structure was produced by running the model with the same seed network and different constant values of preferential attachment and link decay. The colors of the agents represent their social license level, while the direction and color of the links show which way opinion is flowing, and the direction of the opinion being transferred. As noted above, direction here refers to a scale of opinion between positive and negative, which are arbitrary modeling distinctions that are not meant to ascribe value to any specific opinion. In this case, red links are those associated with the transfer of opinion *against* consent for a focal organization, whereas blue links are associated with the transfer of opinion *for* consent.

The ability of a group of stakeholders to grant social license is largely dependent on the amount of "social capital" that is owned by the group as a whole. As defined by Adler and Kwon, social capital is the level of goodwill held by members of a group toward one another (Adler and Kwon 2002). It manifests through solidarity and norm adherence, influence, and access to information, and can be measured through the structure of the network, the quality of relationships, and the level of shared understanding or collective feeling. In an issues-based stakeholder group, this means that the structure of the network (closure/core-periphery), the quality of relationships (link strength), and level of shared understanding (shared opinion) can give some indication of the overall level of social capital owned by the group as a whole. This is critical for social license, because the level of social capital owned by a group will directly affect that group's ability to establish and adhere to norms such as the initial issuance and continued ratification of the social license of either

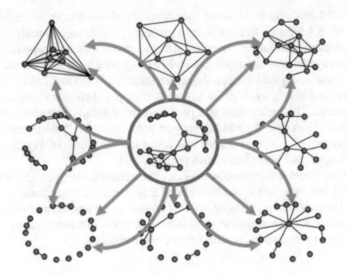

Fig. 5.5 Potential paths of network evolution

internal or external enterprises (Boutilier 2009). What follows is an illustrative interpretation of a translation from network structure to social dynamics. All references to grid position refer to Figs. 5.4 and 5.6.

The equality network (upper left) was produced by running the model without preference for attachment or link decay. The intuitive result is that without constraint on where links can be formed and no mechanism by which they may disappear, the network grows until it is nearly saturated with links. The reason it does not become fully saturated (i.e. density = 1) is because a consensus is reached before all agents have a chance to interact. When all agents reach the same consent level, the signum function from Eq. 5.1 ($sgn(C_i - C_j)$) returns 0, and the model stops since opinion will no longer transfer between agents.

The real-world interpretation of this structure is that because each node is connected to each other node (at least within a geodesic distance of 2), it is easy to establish and enforce norms, including social license to operate. The cohesiveness of the group, however, may lead to a situation in which norms are so well enforced that there is little room for dissent and the network may become a kind of "echo chamber", in which change in group thinking becomes very difficult.

The committee/tribe network (top center) is the result of running the model with no link decay but a high amount of preference for attachment (>90%). No existing links are destroyed as the model runs, but as there is some constraint on the formation of new links, a core-periphery structure begins to emerge in which there is a very slight power-law distribution with the few "core" agents having more links and the many "periphery" agents having fewer. This is not as pronounced as the "dictatorship" network discussed later, as the periphery agents still maintain a high level of connection to the core and to other periphery agents.

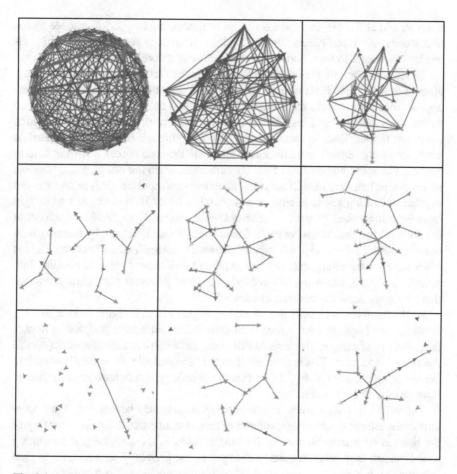

Fig. 5.6 ABM network templates

Norms are fairly easily established in the committee network, but with the emergence of the core it becomes more dependent on the opinion of the initially highly connected agents. The network is capable of issuing a social license, but may be prone to grid-lock and internal political struggles as a result of core agents struggling to represent the competing interests of their well-connected periphery. They can also be prone to cliques and unable to accept new stakeholders as they are activated by a given issue.

An extremely high preference for attachment (>99%) with no link decay produces the "accountable leadership" network in the top right of Fig. 5.6. In this network, there is still a strongly connected core, but the periphery is less well-connected to each other and the power-law distribution of centrality scores is more pronounced. This network can produce and enforce norms fairly easily because the central leadership can efficiently disseminate opinion toward the periphery. It may be kept in check, however, by the competing interests of its periphery connections. It can issue a strong social license, but can also adapt to norms and respond to outside forces and accept new stakeholders. Due to the high preference for attachment, this network takes longer to form than some of the other structures.

The cluster network (middle left) forms from the faction seed when a moderate amount of link decay is introduced with no preference for attachment. As links disappear from the network, the model relies more and more on equivalent-influence transactions and less and less on the normal influence driven transactions, which simulates the small-scale accumulation of power through negotiation of peers in much smaller groups. The more central agents in the seed model, however, tend to become the individual centers of the clusters, even when the bridging and bonding relationships that they held in the core group have disappeared. This network is not capable of issuing social license as a whole, but instead, consent must be sought from each individual grouping, which makes it inherently unstable and subject to the political back-and-forth as each cluster tries to gain influence by bonding with other clusters. In the model, this type of network is somewhat rare to observe and is often merely a transitory state as clusters either break down further or coalesce into a larger structure. More work is needed to explore potential parametric combinations that may stabilize this type of network.

As the more central agents in each cluster begin to form bridging links with other clusters, they begin to form a core with their individual cluster peripheries now at the periphery of a larger structure. At that point, an "exclusive elite" network (middle right) begins to form. This is different than the "accountable leadership" network in the upper right corner of Fig. 5.6 in that the periphery agents have not had time to form links with each other.

The ability of such a group to issue a social license may be tenuous, as the opinions of the elite may not represent those of their constituents. If the directionality of the links is such that the core is dictating opinion to the periphery, the resulting social license may become quickly changed as the periphery agents begin to gain influence by forming links with each other. If the opposite occurs (periphery dictating to core), then the "consensus" represented by the core may be subject to frequent changes as the periphery fight to convince the core of their various opinions.

Real-world networks with no organization (lower left) are usually only encountered after a breakdown of social capital, such as after a war or natural disaster. This type of network is characterized by a complete lack of connection, and an inability to establish or enforce norms. It is produced in the model by having a high rate of decay. It is impossible for a network without structure to enforce norms until after some level of structure begins to appear.

If the "no organization" network has a low preference for attachment and the link decay rate is lowered, it will eventually pass through the "clusters" stage to an "equality" or "committee" type structure. If they start to form connections with a moderate level of attachment preference, balanced with the decay rate, leaders will begin to emerge as previously disconnected agents begin to form links with a single or a few well-connected nodes (lower center). This leader can either become a bridging agent to increase the overall level of social capital held by the group, or enforce norms that ensure that they continue to own nearly all of the social capital themselves, resulting in a dictatorship network (bottom right).

The dictatorship network is characterized by a high level of preference for attachment and a moderate level of link decay. It is reached as either a more well-connected network loses periphery links to decay at the same time that the core is strengthened by the attachment preference, or when a less well connected network begins to form core-periphery clusters that are combined through a preference for attachment to a single actor whose influence has already begun to emerge. This type of network can strongly enforce the opinions of the central node as norms for the group, but are inherently unstable for the lack of periphery connections. If the central node is removed, for example, each resultant cluster becomes a nucleation site for either a new dictator node or, if the preference for attachment is decreased, a more well connected central leadership node such as an elite, committee, or accountable leadership network.

The network structures explored in this chapter were all produced by the agent-based network model, showing the general capabilities of such a method for assessing the ability of a group of agents to issue and maintain a level of social license. As a general rule, increasing preference for attachment leads to an increased core-periphery structure, while increasing the link decay rate decreases the closure structure. Some of the model networks are stable for a given parametric space, while others are transitional states that occur as a given network evolves from one stable state to another.

Concluding Remarks

Agent-based modeling is a powerful tool for reproducing real-world social phenomena such as network evolution based on the diffusion of opinion. Using this method and extending the existing published models with network formation logic, the authors were able to reproduce the structural characteristics found in actual social networks which formed around social license to operate. The fact that the model was able to reproduce the structural features of observed real-world networks so well is a strong validation of the method and modeling algorithms employed in making it. Social license is one example of norm-enforcement behavior that emerges from the collective decision-making of a group of autonomous individuals operating within the framework of influence and opinion. This model could also be applied to problems such as resource allocation, political alliance formation, and many others in which influence is a strong mechanism for opinion diffusion. Capturing the effects of these criteria on the formation of network structures is crucial to understanding how and why social license is granted or withheld, and the implications that opinion diffusion have on the mechanisms of network formation.

References

Adler, P. S., & Kwon, S.-W. (2002). Social capital: Prospects for a new concept. *The Academy of Management Review, 27*(1), 17–40.

Albert, R., & Barabási, A.-L. (2002). Statistical mechanics of complex networks. *Reviews of Modern Physics, 74*(1), 48–94.

Axelrod, R. (1997). The dissemination of culture. *Journal of Conflict Resolution, 41*(2), 203–226.

Bahr, K., & Nakagawa, M. (2017). The effect of bidirectional opinion diffusion on social license to operate. *Environment, Development and Sustainability, 19*(4), 1235–1245.

Barabási, A.-L., & Albert, R. (1999). Emergence of scaling in random networks. *Science, 286*(5439), 509–512.

Borgatti, S. P. (2005). Centrality and network flow. *Social Networks, 27*(1), 55–71.

Boutilier, R. (2009). *Stakeholder politics social capital, sustainable development, and the corporation*. Stanford: Stanford University Press.

Boutilier, R., & Thomson, I. (2009). How to measure the socio-political risk in a project. In *XXVIII Convención Minera Internacional*, pp. 438–444.

Deffuant, G., Neau, D., Amblard, F., & Weisbuch, G. (2000). Mixing beliefs among interacting agents. *Advances in Complex Systems, 3*(01n04), 87–98.

Erdös, P., & Rényi, A. (1959). On random graphs. *Publicationes Mathematicae, 6*, 290–297.

Freeman, R. E., & Mc Vea, J. (1984). *A stakeholder approach to strategic management*. (Working Paper 01-02). Charlottesvill: Darden Graduate School of Business Administration.

Gilbert, N. (2008). *Agent-based models*. Thousand Oaks: Sage Publications Inc.

Hegselmann, R., & Krause, U. (2002). Opinion dynamics and bounded confidence: Models, analysis and simulation. *Journal of Artificial Societies and Social Simulation, 5*(3), 1–33.

Judd, J. S., & Kearns, M. (2008). Behavioral experiments in networked trade. In *Proceedings 9th ACM Conference on Electronic Commerce – EC '08*, p. 150.

Kearns, M. (2012). Experiments in social computation. *Communications of the ACM, 55*(10), 56.

Mitchell, R. K., Agle, B. R., & Wood, D. J. (1997). Toward a theory of stakeholder identification and salience: Defining the principle of who and what really counts. *Academy of Management Review, 22*(4), 853–886.

Nakagawa, M., Bahr, K., & Levy, D. (2012). Scientific understanding of stakeholders' behavior in mining community. *Environment, Development and Sustainability, 15*(2), 497–510.

Okada, I. (2011). An agent-based model of sustainable corporate social responsibility activities. *Journal of Artificial Societies and Social Simulation, 14*(3), 1–29.

Parsons, R., & Moffat, K. (2014). Constructing the meaning of social licence. *Social Epistemology, 28*(3–4), 340–363.

Preusse, J., Kunegis, J., Thimm, M., & Sizov, S. (2014). DecLiNe – Models for decay of links in networks. *arXiv.org*. https://dblp.org/rec/bib/journals/corr/PreusseKTS14; http://arxiv.org/abs/1403.4415

Prno, J., & Slocombe, D. (2012). Exploring the origins of 'social license to operate' in the mining sector: Perspectives from governance and sustainability theories. *Resources Policy, 37*(3), 346–357.

Roberts, S. B. G., & Dunbar, R. I. M. (2015). Managing relationship decay: Network, gender, and contextual effects. *Human Nature, 26*(4), 426–450.

Thomson, I., & Boutilier, R. G. (2011). Social license to operate. In P. Darling (Ed.), *SME mining engineering handbook* (3rd ed., pp. 1779–1796). Englewood: Society for Mining, Metallurgy, and Exploration, Inc.

Watts, D. J., & Strogatz, S. H. (1998). Collective dynamics of 'small-world' networks. *Nature, 393*(6684), 440–442.

Chapter 6
Remarks on the Convergence of Opinion Formation in the Presence of Self-Interest and Conformity

Emiliano Alvarez and Juan Gabriel Brida

Introduction

Economic and social systems can be understood and studied from the relationships and interactions between its various components. In particular, it is interesting to know the dynamics of social decision processes from a bottom-up (Tesfatsion 2002) approach, where from individuals who do not know the complexity of the system, emerging patterns of the process occur. In order to explain the processes of decision making by agents, it is necessary to start from models that take these particularities into account.

In this paper we show the ways in which the action of an agent conditions the rest, assuming that the agents have intrinsic preferences over the actions to be taken. A dynamic version of Conley and Wooders (2001) was studied, where the solutions to the model were based on different assumptions about the behavior of the agents and their interactions. This is a different approach to the issue of intertemporal change in decision-making processes, with differentiated individuals, cognitive and processing constraints, and decisions that change over time according to the mechanisms of information propagation. It integrates the study of complex systems, the evolutionary dynamics and currents of psychology and cognitive sciences, to achieve the desired ends.

In this decision model, agents must decide between the available options, starting from a utility function where the options chosen by the closest individuals are taken into account. Coalitions between individuals in groups (an emergent feature of the model) arise from the iterative decision-making process, based on their individual preferences (unobservable characteristics) and actions taken by other

E. Alvarez (✉) · J. G. Brida
GIDE-DMMC, Facultad de Ciencias Económicas y de Administración,
Universidad de la República, Montevideo, Uruguay
e-mail: ealvarez@ccee.edu.uy; gbrida@ccee.edu.uy

© Springer Nature Switzerland AG 2019
D. Payne et al. (eds.), *Social Simulation for a Digital Society*, Springer
Proceedings in Complexity, https://doi.org/10.1007/978-3-030-30298-6_6

individuals (observable characteristics). When we take into account both aspects, the models formulated are similar to those of Brida et al. (2011) and Alvarez and Brida (2019).

In this paper we focus on two fundamental aspects. First, whether the system converges to a stable situation. Second, what type of solutions are reached, i.e., studying the sensitivity of the results to the parameters of the model.

We must also define the way in which we measure the speed of convergence of the model. In statistical physics, a measure often used is the interface density (ρ). Following Castellano et al. (2009), we can define interface density as the proportion of nearest pairs with a different state. A consensus scenario implies $\rho = 0$ and a disorder state implies $\rho =$ (options $-$ 1/options), when options ≥ 2. The results show that for the values of the parameters analyzed, in each simulation the average value of ρ decreases. However, the value reached and the speed depends on the initial setting.

Another aspect studied is whether this decrease of ρ follows a power-law, like many other phenomena in social sciences (for examples of other power-law in economics; see Gabaix 2009, 2016). The value of α is analyzed and compared with the canonical models of social preference.

This paper is organized as follows. Section "Research Design" introduces the model, following the ODD Protocol (Grimm et al., 2006). Here, the model will be explained in greater detail, as well as the working hypotheses and the experimental design. Section "Results" describes the empirical results, presenting and discussing the simulations. Finally, we present the conclusions, limitations of the study and future research.

Research Design

We develop the model proposed in the previous chapter as an agent-based model in order to be able to simulate interactions at the local level and see what effects they produce at the global level. The NetLogo BehaviorSpace tool was used as it allows to carry out the simulations by exploring the parameter space of the model and the R package to work with the model outputs. In each period we can know the decision of each agent, the number of groups formed and the number of agents that change their decision. For documentation and description, we used the ODD Protocol, version 2 (Grimm et al. 2010).

Model Description

Purpose: The goal of this model is to analyze the decision processes of individuals and their dynamics. It is important to know if differences in the relevance of individual and social preferences give rise to the emergence of consensus or the formation of groups.

Entities, State Variables, and Scales: The model consists of individuals, located in cells of a two-dimensional grid. They interact with others through the public information they provide via their chosen option d ($d = 1, \ldots, D$). In this context, there are D different options.

Process Overview and Scheduling: In the base model, at each period of time, agents interact with a subset of the whole population – a Moore neighborhood with radius 1. In order to choose between different options available, individuals consider which of the alternatives allows to maximize utility considering three factors: (i) its preferences on each option, (ii) the option selected in the previous period and (iii) the option selected by the individuals of its reference group. Then, we compare this base model with a variant of this, in which individuals randomly choose the option with a probability *mut*.

Design Concepts: This model analyzes the dynamics arising from the interaction between individuals. Following Conley and Wooders (2001) and Brida et al. (2011), the individual utility function is made up from individual and social preferences. The formation of groups of individuals with similar choices is an emerging feature of the system. This process can lead to unanimity (global convergence), defined groups (local convergence and global polarization; see Axelrod 1997) or without defined patterns (global and local fragmentation). The individual utility function $u_{i,t}$ can be written as:

$$u_{i,t}\left(d, m_{d,t}\right) = \frac{a}{D} * f_i(d) + b * m_{d,t} - c * \left(1 - m_{d',t}\right) \tag{6.1}$$

where:

- $f(d)$ are individual preferences over the set of different options, exogenous to the model.
- $m_{d,t}$ is the proportion of individuals who choose d in period t in the neighborhood of i. $m_{d',t}$ is the proportion of individuals who does not choose d in period t in the neighborhood of i.
- D is the number of crowding types available.
- a, b and c are the parameters that will allow us to study different behaviors.

Individuals only know their preferences about the set of options, their decision respecting which option they have chosen in the previous period and the decisions taken in their reference group. Then, interactions occur from available public information about the option selected.

Observation: From the simulation rounds, the following information is obtained for each time period:

- Number of groups formed, differentiated between those with one individual and those with several individuals.

- Proportion of individuals that opted for each option.
- Proportion of individuals that change their option, compared to the previous period.
- Average of ρ.

As discussed above, the option preferred by each agent, as well as parameters a, b and c are randomly generated in each simulation. As a result, the initial conditions vary between simulations. Model parameters are characterized in Table 6.1.

We perform 500 replicates of each simulation, by using: four values of *mut* (0, 0.005, 0.01 and 0.015) and two, three or four available options.

Results

This section uses a series of computational experiments in order to analyze the emergent properties of the model. In particular, we study the convergence of the base model and propose some variations. The sensitivity of the results to model parameters is examinated and the speed of convergence is analyzed. This section concludes with a comparison with other relevant social choice models.

Convergence

As stated previously, the system converges towards its equilibrium situation when the density interface or the number of groups is stabilized and reaches a minimum. Defining group as the individuals of the grid who share the same chosen option and who are in a neighborhood of radius 1, the interaction process converges to a state with a lower number of groups.

Table 6.1 Model parameters

Parameter	Description	Value
N	Total of individuals	$35 \times 35 = 1225$
t	Time periods	800
$f(d)$	Individual preferences	$U \sim [0, 1]$
a	Individual utility parameter	$U \sim [0, 1]$
b	Social utility parameter	$U \sim [0, 1]$
c	Antisocial utility parameter	$U \sim [0, 2]$
d	Crowding type	$d \in \{1,\dots,D\} = \mathcal{D}$
mut	Mutation rate	$[0, 0.005, 0.01, 0.015]$

Figure 6.1 shows that for different settings, the number of groups decreases over time. Figure 6.1a compares the base model – whose individuals have two options available – with a model where individuals have a propensity to randomly change the chosen option in a proportion *mut*. One aspect to be highlighted is that in the models with random mutations, the quantity of groups is smaller; however, we should note that the number of groups is significantly smaller when *mut* = 0.5% compared to the model with *mut* = 1.5%. Therefore, the relationship is not linear between the mutation probability and the quantity of groups reached. Figure 6.1b compares the same model but with a different number of options available – two, three or four options. As the number of options increases, it is observed that the number of groups increases. This result affirms that as the available options increase, the polarization in a larger number of groups is more frequent. It should also be noted that as the available options increase, a greater number of periods must pass until the equilibrium is reached.

Figure 6.2 shows the same behavior in the interface density ρ. It is observed that on average, from a base scenario with two options (see Fig. 6.2a) individuals go from 50% of neighbors with a different choice to 11% on average. The interaction process leads to opting for the same options that the closest individuals take, although with a limit determined by the importance that each individual has of the individual and social utility. However, the presence of mutations reduces this limit and allows the number of individuals with different decisions in the neighborhood to be significantly lower. Figure 6.2b compares the value of ρ found by changing the number of options available. When we increase the options available to three or four, we see that ρ stabilizes at 12.5%. There are no significant differences between three or four options, although their initial value is different (66.6% and 75% respectively).

It is important to know if the tendency to conform, understood as the bias to take the same decisions as those decided by the rest of the group, depends on the parameters of the utility function. A greater tendency to conform implies lower ρ values, where consensus implies $\rho = 0$.

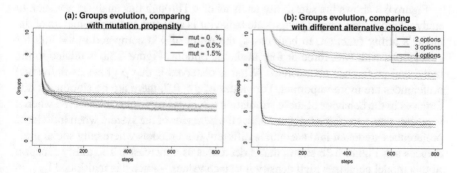

Fig. 6.1 Decrease in the quantity of groups. Notes: Confidence intervals at 95% (dotted lines). (**a**) Comparing the number of groups in the base model with two options and a model whose individuals have mutation propensity. (**b**) Evolution of the quantity of groups, comparing settings with different number of options

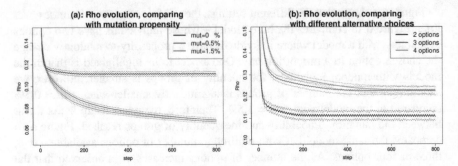

Fig. 6.2 Decrease in ρ. Notes: Confidence intervals at 95% (dotted lines). (**a**) Comparing the number of groups in the base model with two options and a model whose individuals have mutation propensity. (**b**) Evolution of the quantity of groups, comparing settings with a different number of options

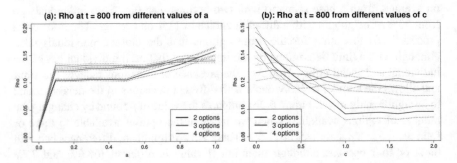

Fig. 6.3 Sensibility analysis of ρ at $t = 800$ according to utility function parameters. Notes: Confidence intervals at 95% (dotted lines). (**a**) Rho values at different a settings (**b**) Rho values at different c settings

Figure 6.3 shows the sensibility analysis of ρ. Through this analysis, we seek to study whether the emerging aggregate behavior is sensitive to the parameters of the model's utility function. In both cases, the result of ρ is compared in $t = 800$ for simulations with two, three or four available options. Figure 6.3a, is related to the individual preference parameter a. What is observed is that ρ grows as individual preferences are more important. For values of $a > 0.7$, there are no significant differences in the behavior of these simulations. A phase transition is observed when a is greater than zero. This fact shows that the behavior of the system when individual preferences are taken into account is different from models where only social preferences act. Figure 6.3b shows that ρ decreases as c increases. The utility function of this model penalizes high density interface values – since c is multiplied by ρ in the utility function, with a negative sign. Therefore, as c grows, it is expected that individuals look for lower values of ρ in order to increase their utility.

Power-Law Distribution of ρ

We want to know now if this behavior follows some kind of law in its temporal evolution. Figure 6.4 shows, in a log-log plot, that the dynamic behavior of the system can be studied from a power-law. The family of power-law functions analyzed are shown in (Eq. 6.1). The speed is determined by the parameter α, which is the slope coefficient in the log-log scale.

$$\rho(t) = \beta + t^{-\alpha} \tag{6.2}$$

We found two different behaviors in Fig. 6.4: on the one hand, the model without mutations (dashed line) and on the other hand the variants of the model with mutations (solid line). The model without mutations allows for the coexistence between options even in the long term: it does not converge asymptotically to a consensus. This result is similar to that found for the Voter Model with dimension > 2 (Castellano et al. 2009) or Schelling's model (Schelling 1971) as well as in small-world networks (Watts 1999), whereas in models with mutations, it converges asymptotically to consensus. This result is innovative, insofar as it is the mutations that allow us to go from a polarized phase to a consensus phase. The speed of convergence (determined by α) is increasing as the number of available options increases.

In Table 6.2 it is appreciated that the parameter α is increasing with the amount of available options, as well as the behaviors of the models with different amount of available options are significantly different.

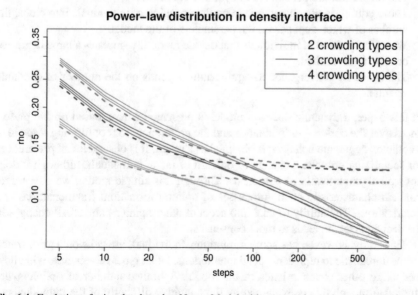

Fig. 6.4 Evolution of ρ in a log-log plot. Notes: Model without mutations (dashed line) and with mutations (solid line)

Table 6.2 Estimation of parameter α of the power-law distribution

Options	Mutation rate (%)	α	CI 2.5%	CI 97.5%
2	0.0	0.10	0.05	0.15
2	0.5	0.12	0.12	0.13
2	1.0	0.11	0.11	0.11
2	1.5	0.12	0.12	0.12
3	0.0	0.15	0.12	0.19
3	0.5	0.14	0.14	0.14
3	1.0	0.19	0.18	0.19
3	1.5	0.17	0.16	0.17
4	0.0	0.20	0.17	0.24
4	0.5	0.20	0.19	0.20
4	1.0	0.20	0.20	0.21
4	1.5	0.20	0.19	0.20

Finally, the α values reached can be compared with other social choice models. In particular, it is similar to that found by Cox and Griffeath (1986) for a voting model of dimension 2, although the speeds found in this model are lower.

Main Results

- The formulated model can reach dynamic equilibria.
- These equilibria are characterized by the grouping of individuals into clubs, the number of which depends on the parameters of the model.
- With a proportion of individuals that decide randomly -mutate- a large degree of consensus is reached.
- The speed of convergence to equilibrium depends on the number of available options.

In this paper, a dynamic decision model is presented where, based on the preferences over the option – individuals – and the preferences for conforming – social –, a complex aggregate behavior is obtained as a result. It is observed that preferences for socializing provoke a change of opinion of many individuals, although it does not lead asymptotically to consensus. Through this simple model, we observe an inherent characteristic of the processes of opinion formation: fragmentations and polarizations. The ability to take into account other opinions and small changes in the decisions taken, leads to more consensus.

This study, however, has some limitations. Individuals are cells on a grid, which seems unrealistic to represent social interactions. Other types of networks will allow evaluating other characteristics of the model. A limited number of options were studied, although this is explained by the cognitive difficulty of the individuals to evaluate and categorize the different options. A model with diffusion of new options is a promising option in this sense.

The lines of future work are linked to jointly establishing a budget constraint. In the presented model, they do not take into account their price or other characteristics. A joint determination model with heterogeneous agents will allow to study the concentration processes in markets from a complexity approach.

References

Alvarez, E., & Brida, J. G. (2019). What about the others? Consensus and equilibria in the presence of self-interest and conformity in social groups. *Physica A: Statistical Mechanics and its Applications, 518*, 285–298.

Axelrod, R. (1997). The dissemination of culture: A model with local convergence and global polarization. *Journal of Conflict Resolution, 41*(2), 203–226.

Brida, J. G., Defesa, M. J., Faias, M., & Pinto, A. (2011). A tourist's choice model. In *Dynamics, Games and Science I* (pp. 159–167). Springer, Berlin, Heidelberg.

Castellano, C., Fortunato, S., & Loreto, V. (2009). Statistical physics of social dynamics. *Reviews of Modern Physics, 81*(2), 591–646.

Conley, J. P., & Wooders, M. H. (2001). Tiebout economies with differential genetic types and endogenously chosen crowding characteristics. *Journal of Economic Theory, 98*(2), 261–294.

Cox, J. T., & Griffeath, D. (1986). Diffusive clustering in the two dimensional voter model. *The Annals of Probability, 14*, 347–370.

Gabaix, X. (2009). Power laws in economics and finance. *Annual Review of Economics, 1*(1), 255–294.

Gabaix, X. (2016). Power laws in economics: An introduction. *Journal of Economic Perspectives, 30*(1), 185–206.

Grimm, V., Berger, U., Bastiansen, F., Eliassen, S., Ginot, V., Giske, J., Goss-Custard, J., Grand, T., Heinz, S., Huse, G., et al. (2006). A standard protocol for describing individual-based and agent-based models. *Ecological Modelling, 198*(1), 115–126.

Grimm, V., Berger, U., DeAngelis, D. L., Polhill, J. G., Giske, J., & Railsback, S. F. (2010). The ODD protocol: A review and first update. *Ecological Modelling, 221*(23), 2760–2768.

Schelling, T. C. (1971). Dynamic models of segregation. *Journal of Mathematical Sociology, 1*(2), 143–186.

Tesfatsion, L. (2002). Agent-based computational economics: Growing economies from the bottom up. *Artificial Life, 8*(1), 55–82.

Watts, D. J. (1999). *Small worlds: The dynamics of networks between order and randomness.* Princeton: Princeton University Press.

Chapter 7
Inequality: Driver or Inhibitor of Collective Action?

Christopher K. Frantz and Amineh Ghorbani

Introduction

Modern societies are characterised by rapid developments in areas including environmental and social awareness, as well as technological development. As a result of this, individual participation in the governance of the society and engagement in its development is once again taking momentum in the form of bottom-up collective action. Collective action provides the opportunity to deal with sustainability and to guarantee the expression of equal and democratic opinion (Chatterton 2016).

In addition, the overt display of socio-economic inequalities, as a side effect of modern societal developments, can also be considered a trigger for collective action movements. Most of those movements either highlight the symptoms of inequality (recall the 1% debate and the associated Occupy movements from 2011 onwards), or drive concrete policy solutions (here the discussion around the universal basic income comes to mind).

Besides being a trigger for collective action, conventional wisdom suggests that inequality has negative effects on the success and durability of collective action (Alesina and La Ferrara 2000; Lijphart 1997). Yet, literature offers a more differentiated picture. For example, Baland and Platteau (2006) identify circumstances in which inequality can act as a driver and inhibitor of collective action. They suggest that influential stakeholders of common-pool resources (CPR) have strong incentives to initiate the management of shared resources in order to preserve the latter (and thus their influence), while less influential stakeholders benefit from shared

C. K. Frantz (✉)
Norwegian University of Science and Technology (NTNU), Gjøvik, Norway
e-mail: cf@christopherfrantz.org

A. Ghorbani
Delft University of Technology (TU Delft), Delft, The Netherlands
e-mail: a.ghorbani@tudelft.nl

© Springer Nature Switzerland AG 2019
D. Payne et al. (eds.), *Social Simulation for a Digital Society*, Springer
Proceedings in Complexity, https://doi.org/10.1007/978-3-030-30298-6_7

governance (and thus equal influence) as a means to prevent overexploitation, and thus to secure their stake.

The question therefore is, if collective action is triggered by inequalities, and acts as a means to promote equal rights and opportunities in a society, can we gain more insight into the circumstances under which inequality can sustain shared governance regimes that are to the benefit of all participants?

This work represents an initial step towards developing an integrated understanding of the influence factors (e.g., resource redistribution preferences, social structures) that drive various types of inequality, including economic inequality in the form of wealth and income, as well as structural inequality based on disparity in social influence.

The structure of the paper is as follows. Section "Inequality and Collective Action" gives some theoretical background on inequality. Section "Model Overview" describes an abstract CPR model that we use to study inequality, followed by an overview of the experimental setup in section "Experimental Setup". Section "Model Results and Evaluation" presents some initial analysis of the simulation results, and finally, section "Experiment 1: No Social Value Orientation, No Segregation" and finally, section "Discussion and Outlook" concludes our findings.

Inequality and Collective Action

The topic of inequality has received attention in a wide area of scientific disciplines, especially since it is seen as a predictor for social disruption and violent conflict, a relationship Lichbach refers to as "economic inequality – political conflict nexus" (Lichbach 1989).

Looking at the drivers of emerging inequality, Tilly (1998) provides a comprehensive long-term overview on the development of inequality, but also explores a broad set of influence factors and social dimensions of inequality (including physical, demographic and economic dimensions). Looking at quantitative evidence, Berman et al. (2016) use empirical data to provide a systematic analysis of policy influence on the dynamics of inequality.

However, when attempting to develop a systematic relationship between both inequality and resulting collective action, the focus on the macro-level perspective can obscure the fundamental micro-level dynamics that are decisive to bring collective action about.[1] In this context, the inequality acts as a seed for disruptive behaviour based on its negative social effects (e.g., reducing the level of social participation and thus development of social capital (Alesina and La Ferrara 2000); undermining democratic processes by increasing the risk of vote-buying (Lijphart 1997)).

---

[1] A convincing account for the case of Rwanda is provided by André and Platteau (1998), who suggest that regional redistribution problems ignited the nation-wide ethnic conflict.

Exploring this relationship further, literature provides a more differentiated picture. Given the documented role of inequality as driver for collective action (both in the positive sense, e.g., establishing resource governance, and negative sense, e.g., violent conflicts), Baland and Platteau (2006) help us identify circumstances in which inequality can act as a driver and inhibitor of collective action in the context of common-pool resource allocation. They suggest that influential stakeholders of common-pool resources have strong incentives to initiate the management of the shared resource in order to preserve it (and thus their influence), while less influential stakeholders benefit from governance as means to prevent overexploitation, and thus secure their share.

This is contrasted with challenges rooted in inequality. Problems arise if the cost of governance (e.g., monitoring) exceeds the benefits drawn from the resource, which, in consequence, can lead to violation of governance commitments by the disadvantaged party. A further challenge arises if the dominating party does not depend on the managed resource (e.g., because of abundant private resources), in which case it can simply exploit the common resource without concern of long-term subsistence. High levels of inequality further drive the risk of bargaining, thus leading to inefficient resource allocation (Bardhan 2005) and concentration of political influence (North et al. 2009).

Looking at this brief overview of related literature, we find that a clear-cut account on the relationship between inequality and collective action processes is hard to establish. To integrate selected documented accounts, we aim to develop an initial model that reflects some of the discussed characteristics to provide a basis for the systematic study of the influence of inequality on the durability and success of collective action.

Model Overview

Given the well-established literature on collective action around common-pool resources, we build an abstract model of a common-pool resource management system to study the relationship between inequality and collective action. At this stage, the model primarily concentrates on wealth inequality (as opposed to income inequality), since it offers a more accurate reflection of socio-economic reality (see e.g., Keister and Moller 2000). In addition, we explore the impact of social influence differences (represented as a tendency to imitate successful individuals' strategies (Bandura 1977)).

The model consists of one shared resource (which has an initial amount and a growth rate) and a collection of agents who appropriate from and contribute to the resource. For this exploration, the resource has a logistic growth rate, such as found in the context of natural resources (e.g., forestry). The agents in the simulation are heterogeneous in terms of initial wealth, social value orientation (SVO) (Griesinger and Livingston 1973), the social influence they have on others, as well as productivity, all of which are operationalised as described below:

- Wealth: Agents are initialised with values drawn from a given distribution. Besides appropriating from the resource, which adds to their wealth, the agents also contribute to it. The level of contribution is dependent on their SVO and the behaviour of their neighbours.
- Social Value Orientation: To model agents' social values with respect to cooperation, we model altruistic, competitive, individualistic and cooperative orientations (Griesinger and Livingston 1973). The proposed model operationalises these as probability ranges that determine an agent's preferred value distribution between itself and others (similar to Murphy et al. (2011)). An agent's orientation is chosen at the beginning of the simulation in the form of a distribution ratio with extreme values that favour altruistic redistribution (Value: 0) or selfish behaviour (Value: 1). It is used to define an agent's resource contribution and appropriation behaviour.
- Social Influence: As an alternative to operating based on their own social value preferences, agents can copy the behaviour of their successful and influential neighbours. The social influence of an agent increases every time others copy its behaviour.

Throughout the simulation, the agents operate in a static spatial environment with a randomly initialised maximum vision radius (*visionRadius*). In a later variation of the model, agents are segregated into dynamic clusters (i.e., neighbourhoods) based on their relative level of contribution and appropriation (referred to as *contribution-appropriation ratio*). This segregation influences the social influence patterns of agents, since they would only copy others in their own cluster (social proximity), rather than the population as a whole. The clustering is performed using the density-based DBSCAN algorithm (Ester et al. 1996).

The detailed agent execution is as follows: After the initialisation with wealth and social value orientation at the beginning of each round as well as random placement within the spatial environment, agents use a parameterised probability (*random action probability*) to determine their resource appropriation based on a random value between 0 and a parameterised maximum appropriation value, which is multiplied by the agent's SVO ratio. Agents then make use of their overall wealth based on a productivity factor that can both have negative and positive values, reflecting both inefficient and efficient use of resources. As an alternative to the autonomous behaviour based on the random action probability, agents identify neighbours in their spatial environment. They copy the appropriation behaviour of the neighbour with the highest combination of wealth and social influence within their vision radius. In this case, the social influence metric of the copied agent is incremented. Following this, a fixed value of wealth (10 units) is deducted to emulate consumption behaviour. Finally, the agent contributes a fraction of its wealth back to the common pool. Equivalent to the decision-making in the case of appropriation, the agent either acts autonomously (based on the random action probability), return a value between 0 and maximum contribution weighted by (1 − *svoRatio*), or by copying the most successful neighbour in their environment.

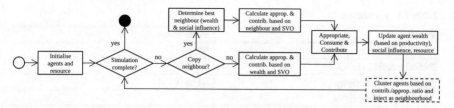

Fig. 7.1 Model overview

As mentioned before, variations of this cycle include the consideration of a dynamic environment, in which neighbours are not determined based on spatial proximity, but rather based on the similarity of their contribution-appropriation ratio determined at the end of each round. The simulation model is outlined in Fig. 7.1 in flow chart notation, with dashed boxes indicating the scenario-dependent additional activity.

Experimental Setup

The model behaviour has been explored through systematic exploration of the parameter ranges. The ranges were chosen based on manual identification of sensible value boundaries. The model parameters and the explored ranges are shown in Table 7.1. Exploration occurred across 600 individual parameter configurations for six scenarios (normal and beta wealth distributions, each with and without consideration of social value orientation and clustering). Each individual simulation configuration is run for 3000 rounds.

Model Results and Evaluation

In order to explore the relation between inequality and collective action in a common-pool resource setting, we have formulated some initial hypotheses which we explored with the model. We observe inequality in two different ways in our experiments: in terms of wealth and in terms of social influence. Our goal is to see whether any of these inequalities' distributions have any relation with the state of the CPR system. The state of the system is defined in terms of the well-being of the resource, and the average wealth and its distribution across the individuals in the system.

To study the correlation between the parameters, we used Spearman's ρ^2, significance level 0.01. Significance tests for different initial wealth distributions are

---

[2] We chose Spearman's ρ due to its rank-based operation that offers a robust analysis of normal vs. non-parametric distributions as well as tolerance against outliers.

Table 7.1 Parameters

Parameter	Value range
Number of agents	50–100
Random action probability	0.1–0.2
Minimum productivity	−1 to 0
Maximum productivity	0–1
Reservation outlook	50–250
Resource growth rate	0.25–0.35
Initial resource amount	10,000–50,000
Maximum contribution	30–40
Maximum appropriation	30–40
Minimum vision radius	2–5
Maximum vision radius	Min. radius + (20 − Min. radius)
Minimum initial wealth	100
Maximum initial wealth	1500
Consumption per round	10
Maximum distance (clustering)	0–0.25
Minimum number of cluster members	3

performed using the Mann-Whitney-Wilcoxon test with a confidence level of 0.95. Table 7.2 shows correlations for selected variables for all explored model variations and different initial wealth distributions.[3] In the following subsections, we discuss selected results of relevance for our exploration.

Experiment 1: No Social Value Orientation, No Segregation

In the first experiment, social value orientation was not considered in the agents' decision making. Furthermore, the agents were not segregated into clusters, and therefore, either acted autonomously, or copied the behaviour of the "best" individuals in their vision radius.

Our null hypothesis was that highly skewed wealth distributions (i.e., high inequality) have a positive effect on the state of the resource. We used beta distributions to represent skewed wealth distributions, and contrast those with wealth initialisation based on normal distributions. Our findings for this base model are as follows:

- For a skewed wealth distribution, we can observe a low negative relationship (−0.39) between wealth inequality and the state of the resource. For a normal wealth distribution, in contrast, the correlation is smaller (−0.33).

---

[3] For the sake of brevity, social influence is referred to as influence in the table.

Table 7.2 The correlation values in the table should be aligned at the decimal sign, since the values, especially when involving both negative and positive values, are hard to read and compare

Correlated variables	Normal	Beta	Normal, SVO	Beta, SVO	Normal, SVO, clustering	Beta, SVO, clustering
Resource μ vs. wealth μ	0.46	0.5	0.53	0.47	0.54	0.55
Resource μ vs. wealth gini	−0.33	−0.39	−0.33	−0.29	−0.27	−0.32
Resource σ vs. wealth μ	0.46	0.5	0.53	0.47	0.54	0.55
Influence σ vs. resource μ	0.35	0.4	0.46	0.4	0.5	0.55
Influence σ vs. resource σ	0.35	0.4	0.46	0.4	0.5	0.55
Influence μ vs. resource μ	0.47	0.52	0.49	0.42	0.49	0.51
Influence μ vs. wealth μ	0.83	0.84	0.92	0.91	0.9	0.9
Influence σ vs. wealth gini	−0.52	−0.61	−0.49	−0.52	−0.53	−0.56
Initial resource μ vs. resource μ	0.99	0.99	0.97	0.97	0.97	0.97
SVO σ vs. wealth μ			−0.18	−0.05	−0.09	−0.07
SVO σ vs. wealth gini			−0.07	−0.15	−0.22	−0.18
SVO σ vs. resource μ			−0.05	0.05	0.05	0.06
SVO σ vs. resource σ			−0.05	0.05	0.05	0.06
Individualistic SVO vs. resource μ			0.02	−0.01	0.01	−0.05
Cooperative SVO vs. resource μ			0.08	0	−0.05	−0.02
Competitive SVO vs. resource μ			−0.12	−0.14	−0.05	−0.11
Altruistic SVO vs. resource μ			0.07	0.11	0.03	0.11
Individualistic SVO vs. resource σ			0.02	−0.01	0.01	−0.05
Cooperative SVO vs. resource σ			0.08	0	−0.05	−0.02
Competitive SVO vs. resource σ			−0.12	−0.14	−0.05	−0.11
Altruistic SVO vs. resource σ			0.07	0.11	0.03	0.11
Individualistic SVO vs. influence μ			0.28	0.24	0.2	0.1
Cooperative SVO vs. influence μ			0.12	−0.01	0.1	0.1

(continued)

Table 7.2 (continued)

Correlated variables	Normal	Beta	Normal, SVO	Beta, SVO	Normal, SVO, clustering	Beta, SVO, clustering
Competitive SVO vs. influence μ			0.12	0.11	0.19	0.16
Altruistic SVO vs. influence μ			−0.34	−0.35	−0.44	−0.35
Individualistic SVO vs. influence σ			0.13	0.08	0.29	0.2
Cooperative SVO vs. influence σ			0.08	0.01	0.09	0.09
Competitive SVO vs. influence σ			−0.05	0.01	0.07	0.02
Altruistic SVO vs. influence σ			−0.02	−0.06	−0.23	−0.14
Cluster count vs. wealth μ					0.39	0.43
Cluster count vs. resource μ					0.35	0.43
Cluster count vs. resource σ					0.35	0.43
Cluster count vs. clustered gini					0.43	0.47
Cluster size vs. clustered gini					0.48	0.5
Cluster count vs. non-clustered gini					0.6	0.63
Cluster size vs. non-clustered gini					0.62	0.65

- For a skewed wealth distribution, wealth level has a low positive correlation (0.5) to resource level, which is slightly more pronounced compared to the relationship between wealth and resource levels for normal distributions (0.46).

Observing the relationship between wealth and resulting distribution of social influence, we can see that:

- Social influence inequality has a low positive correlation with resource state, with slightly stronger relationships for beta distributions (0.4) compared to normal distributions (0.35).
- Inequality in social influence furthermore has the same correlation with wealth inequality as for wealth levels (beta: 0.4; normal: 0.35).
- The social influence level has a moderate positive correlation with resource level (normal: 0.52; beta: 0.52).

These initial results show that when the community is not segregated, and when agents make decisions on appropriation and contribution levels without the consideration of their own social values, the unequal distribution of wealth has a negative relationship to the overall resource state. This is the case both for initial distributions based on non-skewed and skewed distributions, but slightly more pronounced for skewed initial distributions. However, given the minimal differences in selected metrics for both distributions, any claims that point to a specific distribution type should be considered with reservation.

Social influence level, in contrast, has a positive relationship to resource metrics, irrespective of the underlying initial wealth distribution, and a very high correlation with wealth levels (which is of little surprise, given its self-reinforcing role in partner selection). Along with this, however, the wider spread of social influence is positively correlated (but to a more moderate extent than influence and wealth level) with wealth equality. As such, some level of diversity in social influence and wealth spread may have a moderating relationship.

Experiment 2: Social Value Orientation, But No Segregation

Extending the focus of the base model, in this experiment agents consider their SVO when making decisions about their appropriation and contribution behaviour.

Exploring the impact of the introduction of SVOs on the configuration of the previous experiment, we could not observe significantly differing correlation values for initial wealth distribution and resulting resource state. However, the systematic stratification of behaviour based on SVOs offers grounds for further exploration avenues. Our first hypothesis was that a highly divergent society in terms of SVO has a positive influence on the wealth distribution of agents and the state of the resource.

Observing the results, diversity in social value orientation did have a low negative relationship to wealth level for normally-distributed initial wealth (-0.18). For beta wealth distributions, in contrast, we could observe that SVO diversity is weakly related to wealth equality (0.15). Although the reported figures are low, we see that the introduction of SVOs emphasises the role of the initial wealth distribution, which had been rather limited in the first experiment.

While SVO diversity did not render conclusive insights with respect to the resource state, selected SVO components (i.e., altruistic, cooperative, individualistic, competitive) did. The state and variation of the resource is weakly negatively correlated with increasing the fraction of competitive agents (normal: -0.12; beta: -0.24). More significant, however, is the relationship of the fraction of individualistic agents and the mean level of social influence (normal: 0.28; beta: 0.24), thus driving the establishment of influence structures based on the higher fraction of individually appropriated resource. Complementing this observation is the relationship between the fraction of altruistic agents and social influence levels; in our model altruism limits the emergence of a social influence structure.

In contrast to the first experiment, an interesting general observation is that the variation of social value orientations appears to have more pronounced effects on social configurations that have non-skewed (here: normal) wealth distributions. As such, more balanced wealth distributions appear more sensitive to diversity of social behaviour, suggesting that skewed distributions may be more robust against significant shifts in cooperation behaviour.

Experiment 3: Social Value Orientation and Clustering (Segregation)

In the third set of experiments, our goal was to see if the segregation of a society based on their ratio of contribution and appropriation affects metrics of well-being in the system (wealth levels, resource level). Apart from the focus on stratified contribution behaviour (modelled using SVOs), the iterative use of clustering represents dynamic social structures based on changing group relationships – in contrast to the static neighbourhood configurations of the previous experiments.

A central hypothesis is that the introduction of neighbourhood clusters has a positive impact on wealth levels and distribution, as well as the resource state. As with all previous experiments, we further explore how clustering interacts with varying initial wealth distributions. So far, we have observed the following:

- Introducing segregation itself has a minor positive relationship with overall wealth levels, irrespective of the underlying initial wealth distribution. Looking at the results in more detail, we can observe that the number of clusters has a moderate correlation with wealth levels (normal: 0.39; beta: 0.43).
- While clustering itself, and the number of clusters specifically, appear related to stronger levels of inequality, the impact varies for individuals that are members of any cluster (normal: 0.43; beta: 0.47) and for individuals that are not a member of any cluster (normal: 0.6; beta: 0.63).
- In contrast to earlier experiments, the diversity of social value orientation has a stronger influence on the resulting wealth inequality (normal: −0.22; beta: −0.18).
- These findings, specifically with respect to clustering, the differentiation between clustered and unclustered agents, as well as social value orientation, suggest that clustering per se introduces greater inequality. However, while overall inequality increases, the relative homogeneity of individuals within clusters may moderate this effect, especially when compared to all remaining unclustered agents. When combined with clustering, the influence of diversity in social value orientation appears to become more pronounced: Since agents' cooperation behaviours are characterised by their respective social value orientation, their economic stratification will be influenced by the initial SVO, and consequently, drive the formation of economically homogeneous clusters.

Overall, these findings suggest that clusters of homogeneous self-reinforcing behaviour based on SVO can lead to an overall improvement in societal well-being, both in terms of wealth as well as resource level. The flip side is that the combined use of

SVOs and clustering manifests the overall wealth inequality in situations of unequal wealth distribution. The correlations between resource level and social influence further suggest the stronger role of social influence structures. However, whether this occurs within or outside of clusters is inconclusive at this stage.

To arrive at firm conclusions, this model still requires further exploration, involving questions such as: What is a desirable number and size of clusters to manage the trade-off between wealth levels and inequality? Furthermore, how do the individual clusters differ structurally? Specifically this latter question requires the analysis of cluster-specific characteristics (e.g., size, wealth, social influence levels and distributions).

Discussion and Outlook

This paper presents an initial step in our research on the role of inequality on the sustainability of collective action. We built a theoretical agent-based model, which is based on theories of CPR systems to study the correlation between inequality and well-being of the system (see Ghorbani and Bravo (2016) for details of a similar model). In addition to wealth and social influence as behavioural determinants, we instantiated agents with varying social value orientations and introduced segregation based on socio-economic variables to emulate the emergence of social structure. The model is further instantiated using different distribution functions that represent both balanced wealth distribution and skewed wealth distribution in order to analyse the impact of initial wealth distributions on the result.

While the first experiment shows that inequality increases with increasing resource and wealth levels (which bears limited surprise), it also shows that an increase in resource level is linked to stratification of social influence. In the second experiment, we explored the impact of social value orientation on resource, wealth and social influence metrics.

The results show that diversity in social value orientation is weakly associated with more balanced final wealth distributions. However, focusing on individual SVO components, we find that specific social value orientations are more decisive in determining the resulting wealth distribution than diversity per se, especially when involving individualistic agents. Another finding is that behavioural diversity (whatever the specific choice of parameter values) has a stronger impact on social configurations with non-skewed initial wealth distributions, leading to the suggestion that skewed configurations may be more robust against bottom-up dynamics.

The final set of experiments introduces segregation into the model based on density-based clustering using the ratio of contribution and appropriation as socio-economic metric. The results show that clustering introduces higher level of social inequality, but also highlight that this effect is more pronounced for non-clustered agents, in contrast to ones that are organised in clusters. This suggests a moderating effect of clustering for larger societies and larger clusters, an aspect that will be subject of further exploration. Clustering further interacts with social value orientation by affording a segregation of groups with shared characteristics, potentially contributing to the aforementioned homogeneity within clusters. However, to provide more conclusive insights, a detailed investigation of the intra- and intercluster characteristics is needed.

An observation that applies across all experiments is that the initial wealth distribution appears less decisive in determining the final distribution than the behavioural characteristics, such as the injected social values or the emergent social structures based on social influence and clustering. This is particularly observable for cases where different input distributions produced marginal variations in resulting metrics, which is potentially rooted in characteristics of the specific chosen distributions, as opposed to reflecting characteristics of skewed vs. non-skewed distributions more generally.

This leaves us with the opportunity to explore empirically-grounded wealth distributions found in contemporary human societies, in contrast to the idealised normal and beta distributions explored as part of this work. Beyond these analytical refinements, the model does not yet consider monitoring and sanctioning mechanisms found in real societies. Furthermore, we plan to validate our findings by comparing the model input and outcomes with a dataset on common-pool resource institutions.

References

Alesina, A., & La Ferrara, E. (2000). Participation in heterogeneous communities. *Quarterly Journal of Economics, 115*(3), 847–904.

André, C., & Platteau, J.-P. (1998). Land relations under unbearable stress: Rwanda caught in the malthusian trap. *Journal of Economic Behaviour and Organisation, 34*, 1–47.

Baland, J.-M., & Platteau, J.-P. (2006). Collective action on the commons: The role of inequality. In J.-M. Baland, P. Bardhan, & S. Bowles (Eds.), *Inequality, cooperation, and environmental sustainability* (pp. 10–35). Princeton: Princeton University Press.

Bandura, A. (1977). *Social learning theory*. New York: General Learning Press.

Bardhan, P. (2005). *Scarcity, conflicts, and cooperation*. Cambridge: MIT Press.

Berman, Y., Ben-Jacob, E., & Shapira, Y. (2016). The dynamics of wealth inequality and the effect of income distribution. *PLoS One, 11*(4), 1–19.

Chatterton, P. (2016). Building transitions to post-capitalist urban commons. *Transactions of the Institute of British Geographers, 41*(4), 403–415.

Ester, M., Kriegel, H.-P., Sander, J., & Xu, X. (1996). A density-based algorithm for discovering clusters in large spatial databases with noise. In E. Simoudis, J. Han, & U. Fayyad (Eds.), *Second international conference on knowledge discovery and data mining* (pp. 226–231). Portland, Oregon: AAAI Press.

Ghorbani, A., & Bravo, G. (2016). Managing the commons: A simple model of the emergence of institutions through collective action. *International Journal of the Commons, 10*(1), 200–219.

Griesinger, D. W., & Livingston, J. W. (1973). Toward a model of interpersonal motivation in experimental games. *Behavioral Science, 18*(3), 173–188.

Keister, L. A., & Moller, S. (2000). Wealth inequality in the United States. *Annual Review of Sociology, 26*, 63–81.

Lichbach, M. I. (1989). An evaluation of "does economic inequality breed political conflict?" studies. *World Politics, 41*(4), 431–470.

Lijphart, A. (1997). Unequal participation: Democracy's unresolved dilemma. *American Political Science Review, 91*(1), 1–14.

Murphy, R. O., Ackerman, K. A., & Handgraaf, M. J. J. (2011). Measuring social value orientation. *Judgment and Decision making, 6*(8), 771–781.

North, D. C., Wallis, J. J., & Weingast, B. R. (2009). *Violence and social orders: A conceptual framework for interpreting recorded human history*. New York: Cambridge University Press.

Tilly, C. (1998). *Durable inequality*. Berkeley: University of California Press.

Chapter 8
The Venezuelan System of Potato Production: A Simulation Model to Understand Roots of Deficiencies

Oswaldo Terán, Christophe Sibertin-Blanc, Ravi Rojas, and Liccia Romero

Introduction

As in many parts of the world, in Latin America, and especially in Venezuela, potatoes are an essential component of human food and so the production of potato seeds is a national concern. The region around Mucuchíes town (at the Jose Antonio Rangel municipality[1]) in the Merida Venezuelan region, western of Venezuela, concentrates a large proportion of the national Venezuelan production of potato seeds (about 90%) and potatoes (about 40%), and it features the typical character- istics and difficulties of the potato seed and potato production system in Venezuela.

---

[1] Mucuchies, the capital of the Jose Antonio Rangel Municipality, is an Andean highland town. The municipality comprises a surface of 721 km², and a population of 19.634 inhabitants in 2013. Muchuchies is at 8° 45′ of north latitude, 70° 55′ of west longitude, and 2.980 m over the sea level (the municipality is between 2200 and 4600 m over the sea level), with a mean annual temperature of 10.5 °C (for more about this, see: https://es.wikipedia.org/wiki/Municipio_Rangel).

O. Terán (✉)
Escuela de Ciencias Empresariales, Universidad Católica del Norte, Coquimbo, Chile

Centro de Simulación y Modelos (CESIMO), Universidad de Los Andes, Mérida, Venezuela
e-mail: oswaldo.teran@ucn.cl

C. Sibertin-Blanc
Institut de Recherche en Informatique de Toulouse (IRIT), Université Toulouse 1 – Capitole, Toulouse, France
e-mail: sibertin@ut-capitole.fr

R. Rojas
Comunidad Mano a Mano, Mérida, Venezuela

L. Romero
Instituto de Ciencias Ambientales y Ecológicas (ICAE), Universidad de Los Andes, Mérida, Venezuela
e-mail: romero@ula.ve

© Springer Nature Switzerland AG 2019
D. Payne et al. (eds.), *Social Simulation for a Digital Society*, Springer
Proceedings in Complexity, https://doi.org/10.1007/978-3-030-30298-6_8

More precisely, the actual Venezuelan system osf potato and potato seed production (SPP) suffers from scarcity and high price of potato seeds; widespread use of low quality potato seeds; too low diversity of potato species in use; appropriation of parts of the profits of the potato and potato seed market by some traders to the detriment of producers and retailers; and contradictory action from the Venezuelan State, which indicates a low level of concern about a fair SPP (in particular, the State's components responsible for Agricultural issues, such as the Ministry of Agriculture and Lands). This state of affairs is far from being a *fair SPP* that would ensure a steady production that meets the national demand for quality and quantity at just/fair prices for producers, distributors and consumers.

Having these two contrasting configurations in mind, we propose two hypotheses about the causes of the actual SPP deficiencies: (I) most public servants are not doing their best for a fair SPP, at the expense of achieving the SPP goals; and, (II) the SPP has structural features that prevent it from moving out of its current configuration and reducing its inefficiency, even if the positive commitment (altruism) of the main actors increases.

Agent-based computational sociology (Squazzoni 2012) offers powerful tools to investigate such hypotheses by the formal modelling of social systems, the analytical study of structural properties and the computation, by simulation, of behaviours likely to emerge from this structure (Axelrod 1997).

In this line, the SocLab framework (Sibertin-Blanc and El Gemayel 2013; El-Gemayel 2013) proposes a formalisation of the Sociology of Organized Action (Crozier 1964; Crozier and Friedberg 1980), which studies how social organizations are regularized, as a result of counterbalancing processes of power relationships between social actors. Power relationships are founded on the mastering by every social actor of some resources that are somehow needed by others. According to this theory, the behaviour of each actor is strategic while being framed by a bounded rationality (Simon 1982). The interaction context defines a *social game*, where every actor adjusts his behaviour with regard to others in order, as a meta-objective, to obtain a satisfying level of capability to reach his goals, a mix of the aims assigned to his role within the organization and of his particular personal objectives. The end of a social game is to reach a stationary state, that is to say a configuration where no actor changes his behaviour, because the current state of the game gives him an ability to achieve the goals that satisfies him. The organization is then in a regularized configuration and can sustainably function in this way. Soclab has already successfully been used to represent and understand better diverse social and organizational settings (Sibertin-Blanc et al. 2013; Sibertin-Blanc and Terán 2014; Terán et al. 2015a).

Our presentation of the growing of potatoes and potato seeds, the potato market and the difficulties of the SPP are based on several papers that report previous participatory research in various potato and potato seed production communities, notably in the Mucuchíes area (e.g., Velásquez 2001; Romero 2005; Romero and Monasterio 2005; Romero and Romero 2007; Llambí 2012; Rojas 2015; Alarcón 2015). These works document and study the actors of the SPP, their objectives, strategies and interactions, as well as the functioning of the SPP, its difficulties, their origins and consequences in relation with the behaviour of the actors. They also show data, for instance, about production of potato seeds, amount of imported seeds, capacity of potato seed warehouses, etc. The SPP has been the subject of Romero's M.Sc., Ph.D., and several research projects. The integral model and simulation results can be found at https://www.openabm.org/model/4606 (Terán et al. 2015b). To determine the concrete elements for the SocLab model, the methodology suggested in Sibertin-Blanc et al. (2013) was followed – e.g., expert domains were consulted.

The paper is organized as follows. The second section introduces questions that motivate this study and how the SPP operates. The third section presents the SocLab model. Section four combines structural properties of the model with simulation results to consolidate the validity of the model, elucidates its steadiness and proposes a way for the (Venezuelan) State to improve the fairness of the Venezuelan SPP. Finally, section five offers a discussion about the consequences of the validity of the hypotheses, and some conclusions.

The Venezuelan System of Production of Potatoes and Potato Seeds

The SPP is a complex system where economic, agricultural and social issues interfere. Let us start indicating the main problems of the system that motivate our study.

Scarcity and high price of consumption potatoes in the market: The low satisfaction of consumers with respect to the availability and price of potatoes is revealed by their turning to alternatives to potatoes such as cassava or yucca (*Manihot esculenta*), among other kinds of foods. However, this substitution does not affect the demand of potatoes because supply is always below demand, preventing potato consumers from having a significant influence on the market.

Appropriation of parts of the profit of the production of potatoes by some wholesalers, intermediaries between potato producers and retailers, who control a high proportion of the market: Acting as a monopoly, these wholesalers manipulate the market, e.g., hiding potatoes. Manipulation of the market is unhealthy, because of its negative consequences for the potato consumers, producers and the whole SPP. Concretely, the manipulation of the market by the monopoly increases the transaction costs between producers and consumers, which contributes to the inefficiency of the SPP and reduces incentives for farmers to grow potatoes.

The widespread use of bad quality seeds: Consumption potatoes commonly recycled as seeds; use of imported rotten or unhealthy potato seeds, which cannot be planted and are lost by potato growers or contaminate the land with parasites or fungus diseases that are often costly to eliminate and last for several years.

The Growing of Potatoes and Potato Seeds

The potato growing process is as follows: An initial plant is obtained either from flowers (sexual reproduction) or from manipulations of plants in a laboratory (biotechnological production), the latter being called *vitro-plant;* from such first plants, *pre-basic seeds* are obtained; from pre-basic seeds *basic seeds* are produced; following this sequence, from basic seeds we get *registered seeds,* then *certified seeds,* and finally consumption potatoes. The productivity of seeds decreases with the number of generations. For instance, for the Granola variety it is about as follows: 1/70 at the first generation (1 pre-basic seed generates 70 basic seeds), 1/50 at the second generation and so on. When the productivity of a seed is about 1/10 (usually from certified seeds), it is considered as a seed for the crop of consumption potatoes.

Three cropping campaigns are needed to get certified seeds from pre-basic seeds, which requires 3 years in Venezuelan highlands Andes. Indeed, potato cropping campaigns may take place only from March–April to August–September for climatic and sanitary (blight disease) reasons. This makes certified seeds culture very expensive. In addition, the selling price of certified seeds is similar to the one of consumption potatoes, so that growing certified seeds is less profitable than growing consumption potatoes. It seems that the low price of seeds has cultural roots: potato producers do not appreciate the potential of certified seeds enough and they use to plant cheap recycled potato seeds, even if their productivity is much lower. Moreover, the tracing of potato seeds (those got after pre-basic seeds, such as basic seeds, registered seeds, and so on) is poorly known, because of the bad organization of the SPP. All in all, a part of the potato seeds goes to the consumption potato market before completing the whole cycle, while consumption potatoes are recycled as seeds.

Actors and Resources of the Potato Production System

Figure 8.1 shows the main actors of the SPP and the resources that circulate among them. PROINPA (2017a, b) is a Civil Association of producers of potato seeds of the Venezuelan Andean highlands, or Páramo.[2] It aims at quantitatively and qualitatively improving the national production of potato seeds (including the whole production process and local know-how). PROINPA produces vitro-plants and grows pre-basic seeds that are sold to potato producers.

Potato producers (POTAT_PROD) grow the successive kinds of seeds until certified seeds (CERTIFIED_S) for their own needs, and finally consumption potatoes (POTATOES). They also plant imported seeds (IMPORTED_S). In addition, we observe a *recycling* phenomenon: potatoes for consumption are used as (low quality) seeds. In fact, the recycling of consumption potatoes is the mean method by producers to regulate the (un)availability of (certified and imported good quality) seeds.

Until 2012, the State provided private operators with foreign currencies for the importation of potato seeds. Now, the AGROPATRIA State agency (hereafter

---

[2] We use the classification of the páramo as a "… regional placement in the northern Andes of South America and adjacent southern Central America. The páramo is the ecosystem of the regions above the continuous forest line, yet below the permanent snowline. It is a Neotropical high mountain biome with a vegetation composed mainly of giant rosette plants, shrubs and grasses" (http://en.wikipedia.org/wiki/Páramo).

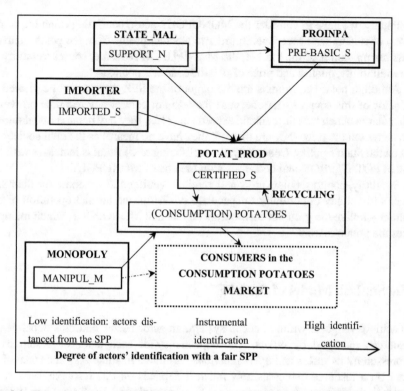

Fig. 8.1 The main actors of the SPP and the resources that circulate among them. Notes: Each actor is shown in a box with bold letter, above an internal box naming the resource(s) it controls. Arrows show the main dependencies of actors on resources. Strongly committed actors are placed at the right side, while lowly identified ones are at the left side

called IMPORTER) (AGROPATRIA 2017) ensures the importation of seeds in addition to the distribution of agro-chemical inputs. This actor is lowly committed with a fair SPP. For instance, it does not care whether imported seeds are finally planted or about their quality: it happens that imported seeds are sold although they have already rotten, or contaminated with fungal diseases or parasites, with severe consequences on the land.

Agricultural affairs are managed by several agencies depending on the Ministry for Science and Technology and mainly the Ministry for Agriculture and Lands (MAL) that designs the agricultural policy. They provide financial, technical, material, training, organizational and other various supports aimed at developing and improving the SPP, but often with some politicization, acting in the service of political parties. Some State agencies (INSAI, SENASEM) are responsible for controlling the quality of potato seeds, preventing diseases and supporting the national production of good quality seeds, but their commitment is poor and their activity lowly effective. These state-dependent departments act in similar ways and may be viewed as a composite but coherent actor, let's call it STATE_MAL, which manages the SUPPORT resource.

Finally, we have to consider the MONOPOLY actor, consisting of an important group of traders who speculate, in order to abstract profits from the potato market. This manipulation of the market, the MANIPUL_M resource, causes volatility in the availability, quality and price of potato seeds and potatoes.

Although potato consumers are the target of the SPP, they cannot be viewed as an actor of this social system because they do not control any resource on which other actors depend: as the demand exceeds the offer, they buy potatoes available on the market at the price they are offered; they have no means, as the civil society, to act on the State's policy. Concretely, the well-being of consumers increases with the one of POTAT_PROD and decreases with the one of MONOPOLY.

Similarly, potato retailers are small vendors, getting a low income for their service, without any opportunity to speculate or disrupt the normal operation of the market – unlike the group of distributors here called MONOPOLY, which manipulates the potato market.

The SocLab Model of the SPP

To achieve its goals within a social system, an actor needs, and thus depends on, resources managed by others. The dependency of each actor on resources is expressed by its *stakes* and *effect functions,* as shown in Table 8.1. The stake of an actor on a resource expresses how much it depends on this resource, on a scale *null = 0, negligible = 1,..., significant = 5,..., critical = 10.* Each actor has 10 stake points to distribute on the set of all resources.

An effect function expresses to what extent the management of a resource (by the actor who controls this resource) impedes or supports an actor in the achievement of its goals. The x-axis corresponds to the state of the resource, *i.e.* how it is managed, from the least (negative direction) to the most (positive direction) cooperative way with regard to all actors. The y-axis corresponds to the actor's capability to achieve its goals, depending on the way the resource is managed.

The impact of a resource r with the management m upon an actor a, *i.e.* the *capability* it provides to this actor to achieve its goals, is the value of the $effect_{r,a}()$ function applied to m, weighted by the stake of a on r. As a result, an actor gets some *satisfaction*, as an aggregation of the impacts it receives from the resources it depends on, and, conversely, it exerts some *influence*, as an aggregation of impacts it grants to actors who depend on the resources it controls. When a system of organized action is in a configuration $m = (m_r)_{r \in R}$, the *satisfaction* got by an actor a is defined as:

$$Satisfaction(a,m) = \sum_{r \in R} stake(a,r) * effect_{r,a}(m_r).$$

Conversely the *influence* exerted by an actor a is defined as the sum of the capabilities it supplies to all actors by the management of the resource(s) it controls, that is:

$$Influence(a,m) = \sum_{r \in R; \, a \, controls \, r} \sum_{b \in A} stake(b,r) * effect_{r,b}(m_r).$$

Table 8.1 Matrix of the dependencies, effect functions and stakes, of actors (in columns) on resources (in rows)

	PROINPA	POTAT_ PROD	STATE_ MAL	MONOP OL	IMPORT ER	Relevance
PREBASIC_S	**4.0**	0.5	1.0	0.0	0.5	6
CERTIFIED_S	1.5	**1.5**	1.5	0.5	1.5	6.5
POTATOES	0.5	**3.0**	3.0	4.0	2.0	12.5
SUPPORT	2.5	2.5	**2.0**	2.0	1.5	10.5
MANIPUL_M	0.5	1.0	0.5	**3.0**	1.0	6
IMPORTED_S	1.0	1.5	2.0	0.5	**3.5**	8.5

Notes: Stake of the actor who controls the relation is in bold character. Each actor allocates a normalized sum of 10 stake marks. For each effect function, the x-axis represents cooperativity in the management of the resource by its controlling actor, while the y-axis is the resulting impediment or facilitation for the dependent actor. The Relevance column indicates the sum of stakes put on each resource

The setting of stakes and effect functions is based on a careful consideration of observations reported in (Velásquez 2001; Romero 2005; Romero and Monasterio 2005; Romero and Romero 2007; Llambí 2012; Rojas 2015; Alarcón 2015). L. Romero has been the main domain expert validating the model presented in this paper. We just sketch justifications of the values of stakes and the shapes of effect functions of the model, a more detailed presentation of the model is given at https://www.openabm.org/model/4606 (Terán et al. 2015b).

PROINPA This actor is highly concerned and positively impacted by its PREBASIC_S own production (stake = 4, linear increasing effect function). Its most important other resources are: the STATE_MAL's SUPPORT (stake 2.5), since it provides technological equipment, pesticides and fertilizers, and financial support; and CERTIFIED_S (stake 1.5), because this determines the demand for PREBASIC_S. The increase of CERTIFIED_S stimulates the demand of PREBASIC_S for the benefit of PROINPA, but not beyond its production capacity. Additionally, PROINPA is positively affected, though indirectly in the production chain, by the production of POTATOES (stake 0.5), whose production needs CERTIFIED_S; and, negatively, by the amount of IMPORTED_S (stake 1), alternative seeds that compete with the national production. High level of MANIPUL_M has a negative, though indirect, effect on PROINPA (stake 0.5), because of its harmful impact on a fair SPP.

POTAT_PROD POTAT_PROD is mainly affected by its POTATOES production (stake 3), and by SUPPORT (stake 2.5) that sustains its activity. It is also concerned, though at a lower degree, by resources required for growing potatoes, *i.e.* CERTIFIED_S and IMPORTED_S (stake 1.5 each). POTAT_PROD gives low importance to MANIPUL_M (stake 1), as it does not make the demand fall below the offer, and to PREBASIC_S (stake 0.5) according to its low interest in cultivating CERTIFIED_S.

STATE_MAL This actor is moderately concerned with the resource SUPPORT it controls, (stake 2) as its existence does not depend on its activity, and its best is to provide a moderate support. STATE_MAL is mainly concerned by the supply of the consumers' demand, POTATOES (stake 3), without worrying too much about their origin and quality. STATE_MAL gives smaller importance to national CERTIFIED_S (stake 1.5) than to IMPORTED_S (stake 2), an easier way to get seeds that sustain potato production. The shape of the effect function expresses that expenses for importation of seeds compete with the budget for supporting the whole SPP. Given the low concern of STATE_MAL with a fair SPP, it does not care much about PREBASIC_S and MANIPUL_M (stakes 1 and 0.5, respectively). For STATE_MAL, the lack of good quality PREBASIC_S seeds is not a problem – it satisfies with recycled or imported potatoes. Also, the effect of MONOPOLY's manipulation on the potato price is not an important concern for STATE_MAL – the availability of potatoes matters more than their price. In addition, to control the MONOPOLY would require a strong and well-organized effort – what is not in the STATE_MAL's agenda.

MONOPOLY The MONOPOLY is highly concerned and directly interested by the resource it controls, MANIPUL_M (stake 3). Regarding other resources, it is interested above all in POTATOES (stake 4), with a medium quantity that provides the better condition for manipulating the potato market. For similar reasons, it is interested in a significant SUPPORT (stake 2) that fosters production of potatoes and, in addition, offers opportunities for profitable loans to buy, *e.g.*, new installations or equipment (trucks to transport potatoes...). MONOPOLY has much less concern with potato seeds, especially with PREBASIC_S, as this resource affects indirectly the level of POTATOES in the market.

IMPORTER IMPORTER is an agency of the Venezuelan State, which is lowly committed in a fair SPP. IMPORTER is interested in the IMPORTED_S resource it controls (stake 3.5), but the effect function reflects that importation of potato seeds is strongly constrained by the availability of foreign currencies. It is also interested (stake 2) in a medium production of POTATOES: if it is too low, there is no need to import seeds, if it is too high, POTAT_PROD's demand of imported seeds will increase, and the poor quality of IMPORTER's work will become more and more obvious. IMPORTER is favored by low levels of SUPPORT (stake 2), and of CERTIFIED_S (stake 1.5). It is also interested in a low level of PREBASIC_S, but weakly, because of its indirect effect (stake 0.5). IMPORTER is somewhat interested in the instability of the national market of POTATOES resulting from MANIPUL_M (stake 1), because the frequent lack of seeds justifies importing potato seeds.

Analysis of the Model

From an analysis of the model, we expect that, on the one hand, enough features of the model are in agreement with the empirical knowledge of the SPP, so that the model may be viewed as a faithful representation of the SPP and, on the other hand, that the formal processing of the model will bring us instructive and useful knowledge on the SPP. In this sense, we address the question: What could be the ways from the actual working of the SPP toward a fair SPP where PROINPA and POTAT_PROD have a high satisfaction and MONOPOLY a low one?

Structural Analysis

We first look at some remarkable configurations of the SPP, which are not socially feasible, but inform us about the potentialities of the structure of the SPP. Table 8.2 presents the state of resources and the satisfaction of actors, in the configurations where the global satisfaction (*i.e.* the sum of all actors' satisfactions) and the satisfaction of each actor are maximized. The configuration for the maximum satisfactions of the whole system (318.0) is closer to the maximal satisfaction of the MONOPOLY (166.9) than any other actor (the next one is the STATE_MAL (136.7)). The best configurations of MONOPOLY and STATE_MAL provide the highest global satisfaction, that is, they realize a compromise between the interests of all others. It would be better if the State plays this role, rather than Monopoly (the best for the SPP is also close to the best for MONOPOLY), an actor strongly

Table 8.2 This table shows (in lines) the configurations (state of resources) and actors' satisfactions for (in columns) maximal satisfactions, global and for each actor

		GLOBAL	PROINPA	POTAT_PROD	STATE_MAL	MONOPOLY	IMPORTER	Fair SPP
State of resources	PREBASIC_S	10.0	10.0	10.0	10.0	-10.0	-10.0	10.0
	CERTIFIED_S	-5.0	-5.0	-7.0	10.0	-3.0	-10.0	-5.0
	IMPORTED_S	-2.0	-10.0	10.0	-5.0	-3.0	-2.0	-2.0
	POTATOES	3.0	10.0	5.0	10.0	-1.0	0.0	3.0
	SUPPORT	4.0	10.0	10.0	-1.0	0.0	-10.0	4.0
	MANIPUL_M	10.0	-10.0	-8.0	-10.0	10.0	10.0	-10.0
Satisfaction of actors	PROINPA	72.1	98.6	64.9	67.5	-24.9	-80.1	81.7
	POTAT_PROD	46.3	59.2	98.6	6.8	5.3	-11.7	66.0
	STATE_MAL	46.4	34.1	-6.2	98.4	6.6	-13.3	56.1
	MONOPOLY	87.4	-51.8	-5.0	-37.3	100.0	75.8	87.4
	IMPORTER	65.9	-13.7	-33.5	1.3	79.9	99.7	45.9
	GLOBAL	**318.0**	126.3	118.7	136.7	166.9	70.4	**337**

Notes: The last column shows the configuration of a fair SPP, when the STATE_MAL adopts energetic measures to prevent speculation on the potatoes market, and it is provided with a 0.5 altruism (see section "Towards a Fair SPP")

distanced from a fair SPP. Even more, the Euclidean distance between the configurations of maximum satisfactions of STATE_MAL and IMPORTER (both part of the Venezuelan State) is larger than their distance with other actors, showing a fragmentation of the behavior of the Venezuelan State.

Certain convergences and conflicts between actors appear: Convergence between PROINPA and POTAT_PROD, on one side, and between MONOPOLY and IMPORTER, on the other side. In both groups, the configuration that maximizes the satisfaction of one of them provides the other with a good satisfaction, and the same holds for their minimum satisfactions (not shown here). A clear conflict between the two groups also appears, since good configurations for one group are bad for the other. It is deceiving to find IMPORTER close to MONOPOLY, an actor distanced from a fair SPP, while far from PROINPA and POTAT_PROD, two key actors for the production of potatoes, the goal of the SPP.

Simulation Results

SocLab includes a simulation engine that computes configurations likely to emerge within the modeled system, by making the actors of the model adapt their behaviors to each other in order to get a good satisfaction, and so play the social game (Sibertin-Blanc and El Gemayel 2013). The simulation algorithm includes some stochasticity so that simulations are run several times. Tables 8.3 and 8.4 show average results of 200 simulations of the SPP model.

The most satisfied actors are MONOPOLY, IMPORTER, and PROINPA, while the least ones are POTAT_PROD and STATE_MAL (Table 8.3). Accordingly, POTAT_PROD and STATE_MAL are the actors having the lowest percentages of satisfaction with regard to the range of their possible satisfactions. This configuration is far from a fair SPP, although the global satisfaction is quite high (307), close to its maximum (318, see Table 8.2).

As Table 8.3 shows, POTAT_PROD has a much higher influence than other actors due to the large amount of stakes on the relation it controls, while MONOPOLY is the actor with the lowest influence. Notice that *STATE_MAL provides only 90.9% of the capability it could give, in accordance with its low commitment for a fair SPP.*

Regarding the state of resources (Table 8.4), results agree with the actual working of the SPP. For instance, high MANIPUL_M corresponds to the usual active disturbance of the potato market by the MONOPOLY. Similarly, PROINPA's production of PREBASIC_S actually is at the maximum of its capacity. Production of POTATOES is in a medium level, much higher than CERTIFIED_S, corresponding to the scarcity of good quality seeds as a result of the low SUPPORT from the State, and its low profitability compared to the production of consumption potatoes. SUPPORT is a bit low, in accordance with the low consideration of Venezuelan State and MAL's for a fair SPP. Importation of potato seeds is low, corresponding to budget limitations.

Table 8.3 The average capabilities provided by each actor (in columns) to each one (in lines) over 200 simulations

	PROINPA	POTAT_PROD	STATE_MAL	MONOPOLY	IMPORTER	Satisfaction	% Satisfaction
PROINPA	39.6	16.8	5.9	−4.8	4.1	61.6	81.3%
POTAT_PROD	5.0	40.5	5.9	−10.0	−6.2	35.2	67.7%
STATE_MAL	9.6	12.3	18.6	−4.8	18.6	54.3	76.9%
MONOPOLY	0.0	32.9	19.7	29.7	5.0	87.3	92.4%
IMPORTER	−4.8	30.7	−1.8	9.9	35.0	68.9	82.9%
Influence	49.4	133.2	48.3	20.0	56.5		
% Influence	99.6%	99.4%	90.9%	99.1%	100%		

Notes: The two last columns give the resulting satisfaction in value and proportion, and the last two rows the corresponding values of influence

Table 8.4 State of resources at the configuration resulting from simulations (average and standard deviation), at the Nash equilibrium and at the configuration of maximum global satisfaction

	Average	Deviation	Nash	Max global satisfaction
PREBASIC_S	9.91	0.14	10.0	10.0
CERTIFIED_S	−4.79	1.01	−7.0	−5.0
IMPORTED_S	−2.2	0.81	−2.0	−2.0
POTATOES	3.64	0.52	5.0	3.0
SUPPORT	1.2	0.69	−1.0	4.0
MANIPUL_M	9.87	0.21	10.0	10.0

Given the low values of CERTIFIED_S and IMPORTED_S, in both the real system and the simulation results, another source of seeds is needed to support the medium production of consumption potatoes. Thus, a high level of recycled consumption potatoes may be inferred from simulation results, which corresponds to the difficulty mentioned above.

All these results are coherent with the problems observed in the SPP, which were numbered above. Let us explain this:

Problem (i). The simulation results show that potato producers are one of the less satisfied actors, while the MONOPOLY is highly satisfied. Remembering that consumers' satisfaction increases with the augmentation of POTAT_PROD's satisfaction and diminution of MONOPOLY's satisfaction, we verify a low satisfaction of potato consumers.

Problem (ii). The high level of MONOPOLY's satisfaction in simulations corresponds to a high achievement of its goal, entailing a high disturbance of the consumption potato market, with high transaction costs for POTAT_PROD and potato consumers, favoring a weak motivation of POTAT_PROD (corroborated by the low value of their satisfaction), what favors, in turn, the deficiency of CERTIFIED_S and increases the price of consumption potatoes.

Problem (iii). The average levels of resources in the simulation configuration (Table 8.4) indicates that the quantities of potato seeds, both imported and certified, are either low or very low, so that another source of seeds is necessary to have the medium level of potato production: the usage of low quality recycled potatoes.

The principal component analysis (see Fig. 8.2), as well as the correlations between actors' satisfactions (not included here for lack of space), show strong positive correlations between the following pairs of actors: POTAT_PROD with PROINPA, and IMPORTER with MONOPOLY, and opposition between these two groups. This is in line with results of the structural analysis above. In addition, opposition also arises between STATE_MAL and the second group (IMPORTER and MONOPOLY). The relation between STATE_MAL and the first group (POTAT_PROD and PROINPA) is ambiguous: there is some convergence regarding component 1, but also opposition regarding component 2. All in all, the three groups are globally equidistant. *STATE_MAL plays an intermediate role, trying to keep a neutral position between the two conflicting pairs of actors. This is not good for a*

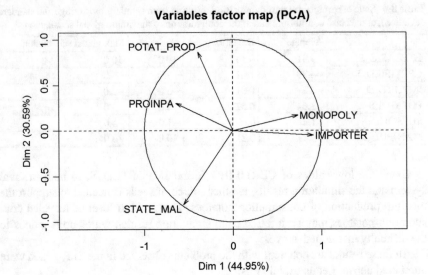

Fig. 8.2 Principal components graph for the actor's satisfaction

fair SPP – a fair SPP would require STATE_MAL to be an unconditional ally of PROINPA and POTAT_PROD.

On the other hand, there is no significant correlation between the state of resources IMPORTED_S and MANIPUL_M, neither between PREBASIC_S and CERTIFIED_S or POTATOES (the only one, though weak, positive correlation is between CERTIFIED_S and IMPORTED). So, *the proximity within each group is just a similarity of interest, which does not give rise to a coordination of their behaviors.* There is no coalition among actors.

In Fig. 8.2, the smaller size of the vectors for PROINPA and MONOPOLY is related to the small deviations of the resource they control (see Table 8.4), compared, *e.g.*, to those of POTATO_PROD and STATE_MAL, and to their small influence (see Table 8.3).

Steadiness of the Actual Configuration

A Nash equilibrium configuration of a social game is one where each actor adopts the behavior that maximizes the benefit for itself. Table 8.4 shows that the states of resources in the simulation configuration are close to both the states at the Nash Equilibrium and to the states at the maximum global satisfaction. Connecting simulation results to structural properties of the model, we observe that the actual behavior of each actor is beneficial not only for itself (close to Nash equilibrium) but also for all the others (close to maximum global satisfaction).

We may conclude from this fact that the actual working of the SPP is very steady, so that it is unlikely that it moves toward a fair SPP without changes in its structure. Indeed, it is conflict-free and its actual working features good performance (the simulation configuration is close to the best global configuration) and no actor has reason to adopt another behavior (the simulation configuration is close to the Nash equilibrium). This fact is confirmed by the low value of the standard deviation of resource states (see Table 8.4): each actor adopts very similar behaviors in all simulation runs, as if it founds no alternative good solution.

Towards a Fair SPP

What if we provided STATE_MAL with a specific rationality that would lead it to adopt the behavior that maximizes global satisfaction (*i.e.*, 4 for the state of SUPPORT) instead of being half-way between this behavior and the Nash equilibrium? In this case, satisfactions of PROINPA and POTATO_PROD increase a little, from 61 to 72 and 35 to 46 respectively, but MONOPOLY still dominates (its satisfaction remains 87).

SocLab offers the possibility to consider "altruistic" actors that are more or less committed in the good working of the whole organizational system and so try more or less to improve not only their own satisfaction, but also the one of all other actors. Altruism can be set between 0 and 1: in case of altruism 0, the default value, the actor worries only about its own satisfaction; in case of altruism 1, it considers only the satisfactions of others (i.e. global satisfaction minus its own). If the STATE_MAL's altruism rises up to 1, simulation results show that the satisfactions of PROINPA and POTATO_PROD increase until 71 and 53 respectively, while the one of STATE_MAL decreases from 54 to 33, and the satisfactions of the two other actors do not significantly change. *However, the levels of resources PREBASIC_S, CERTIFIED_S and POTATOES vary very slightly, and the MONOPOLY is still the more satisfied actor, i.e., the system does not move towards higher satisfaction for consumers, a fair SPP.*

It seems that the way towards a fairer SPP requires changes in the SPP structure, not just in the actors' behaviors. Assume that the Venezuelan State is in capacity to adopt energetic measures to prevent and repress speculations on the potato market. This gives rise, in the model, to reversing the effect function of the MANIPUL_M resource on the MONOPOLY actor as a linear decreasing function, since speculation becomes very costly for traders. Assume also that the STATE_MAL actor is more concerned by the well-operating of the SPP and behaves with altruism of 0.5. Simulation results provide the configuration shown in the last column (at the right) of Table 8.2, which is much fairer than the configuration shown in the first row of this table. The configuration of such a fairer SPP features a higher global satisfaction (337) than the maximal global satisfaction got in the SPP (318). The MONOPOLY, that should be renamed WHOLESALERS, is still the most satisfied actor. This is a very common situation in agricultural markets, especially in developing economies where the small number of intermediaries vis-à-vis many small producers provides the former with advantages in price negotiations.

Discussion and Conclusion

While the simulation model exhibits strong correspondences with the real SPP, several surprising sources of the SPP inefficiency were also found. This new knowledge about the inefficiency of the SPP assists us in concluding about the two hypotheses set in the introduction of the paper:

I. Most State agents are far from making their best in favor of a fair SPP, to the detriment of SPP's goals achievement.
II. The SPP has structural features that prevent the system from moving out of its configuration to improve its effectiveness, even if the positive engagement (altruism) of the main actors increases.

Among the new knowledge given by the analysis of the model, all supporting the hypotheses, we have:

(a) The low capability given by STATE_MAL despite the high relevance of the relation it controls, strengthened by the fact that it provides only 90.9% of the capability it could give, shows its low commitment to a fair SPP.
(b) Unlike its duty, STATE_MAL maintains an ambiguous relationship with the two conflicting pairs, whereas it should be an unconditional ally of POTAT_ PROD and PROINPA.
(c) This is the MONOPOLY's best configuration that provides the highest global satisfaction, realizing a compromise among the interest of all others. That the best for the whole SPP is close to the best for the MONOPOLY, an actor strongly distanced from a fair SPP, means that the balance between the actors is not in favor of a fair SPP.
(d) The convergence between PROINPA and POTAT_PROD, on one side, and between MONOPOLY and IMPORTER, on the other side, and opposition between the two pairs, revealed by both the structural analysis and the principal components analysis of simulation results, indicate that IMPORTER is not interested in a fair SPP.
(e) The large distance between STATE_MAL and IMPORTER found in the structural analysis, and their reciprocal opposition showed by the principal components analysis, demonstrate fragmentation and lack of cohesion in the role of the Venezuelan State – even worst, none of these actors demonstrates appropriate concern with a fair SPP.

The steadiness of the regulated state (it is close to the Nash equilibrium and to the Max Global satisfaction state) makes very hard to overcome these difficulties. Even worst, varying the in-group identification (altruism) of key actors such as

STATE_MAL does not generate changes in the management of the resources PREBASIC_S, CERTIFIED_S and POTATOES, keeps MONOPOLY as the most satisfied actor, *i.e.*, and does not move the system towards more satisfaction for consumers, or a fair SPP.

Thus, the only way to modify the system behavior towards a fair SPP is by structural changes producing another social game. The computational modeling and simulation allows us to test a scenario that shows interesting results: if the State strengthens the control over the potatoes market, by taking vigorous measures to prevent MONOPOLY's speculation and also increasing its responsibility, a much fairer SPP arises (for instance, the global satisfaction is higher than the maximal global satisfaction in the current SPP, and the MONOPOLY dominance over the game strongly decreases).

Further research will be directed towards investigating other scenarios for a fair SPP involving structural changes, in order to understand better how the SPP could be transformed. In SocLab, this implies a re-definition of actors' stakes and effect functions, corresponding to changes in the organization and actors' representations in the real system.

References

AGROPATRIA. (2017). *AgroVenezuela*. Ministerio del Poder Popular para la Agricultura y Tierra. Retrieved on March 2017 from: http://www.agropatria.com.ve/

Alarcón, L. (2015). *Modelado y Simulación del Sistema de Acción Organizada de Producción de Semilla de Papa del Municipio José Antonio Rangel*. Undergraduate Final Project. Universidad de Los Andes, Mérida.

Axelrod, R. (1997). Advancing the art of simulation in the social sciences. In R. Conte, R. Hegselmann, & P. Terna (Eds), *Simulating Social Phenomena* (pp. 21–40). Lecture Notes in Economics and Mathematical System 456. Springer-Verlag.

Crozier, M. (1964). *The bureaucratic phenomenon*. Chicago: University of Chicago Press.

Crozier, M., & Friedberg, E. (1980). *Actors and systems: The politics of collective action*. Chicago: The University of Chicago Press.

El-Gemayel, J. (2013). *Modèles de la rationalité des acteurs sociaux* (PhD of the Toulouse University). Toulouse University, Toulouse, France.

Llambí, L. (2012). *Procesos de Transformación Territorial y Agendas de Desarrollo Rural: El Municipio Rangel y la Asociación de Productores Integrales del Páramo (PROINPA) en los Andes Venezolanos*. Agroalimentaria, Mérida, pp. 19–30.

PROINPA. (2017a). *Red Socialista de Innovación Productiva Integral del Cultivo de Papa municipio Rangel Estado Merida, Venezuela*. Retrieved on March 2017 from: http://www.innovaven.org/Timotes/2a%20Sesi%C3%B3n%204a%20R%20Romero.pdf

PROINPA. (2017b). *PROINPA*. Mérida, Venezuela. Retrieved on March 2017 from: http://proinpameridavenezuela.blogspot.com/

Rojas, R. (2015). *Descripción del Sistema de Acción Organizada de Producción en Torno al Cultivo y Producción Semilla de Papa: Caso Municipio José Antonio Rangel*. Undergraduate Final Project. Universidad de Los Andes, Mérida.

Romero, L. (2005). *La estrategia de la semilla en el sistema papero de los Andes de Mérida. Una visión desde la perspectiva agroecológica* (Tesis de Doctorado). Instituto de Ciencias Ambientales y Ecológicas, Universidad de los Andes, Mérida, Venezuela.

Romero, L., & Monasterio, M. (2005). *Semilla, Actores e Incertidumbres en la Producción*. Papera de los Andes de Mérida. Realidades y Escenarios Bajo en Contexto Político Vigente. Cayapa, pp. 35–58.

Romero, L., & Romero, R. (2007). *Agroecología en los Andes Venezolanos*. Investigación, pp. 52–57.

Sibertin-Blanc, C., & El Gemayel, J. (2013). Boundedly rational agents playing the social actors game – How to reach cooperation. In L. O'Conner (Ed.), *Proceedings of the IEEE Intelligent Agent Technology Conference*. Atlanta, GA, 17–20 November 2013.

Sibertin-Blanc, C., Roggero, P., Adreit, F., Baldet, B., Chapron, P., El Gemayel, J., Mailliard, M., & Sandri, S. (2013). SocLab: A framework for the modeling, simulation and analysis of power in social organizations. *Journal of Artificial Societies and Social Simulation, 16*(4), 8. Retrieved on March 2017 from: jasss.soc.surrey.ac.uk/16/4/8.html

Sibertin-Blanc, C., & Terán, O. (2014). The efficiency of organizational withdrawal vs commitment. In F. J. Miguel (Ed.), *Proceedings social simulation conference (SSC 2014), Barcelon (Spain)*, Universitat Autonoma de Barcelona, Barcelona, 2–5 September 2014.

Simon, H. A. (1982). *Models of bounded rationality: Behavioral economics and business organization (Vol 1 and 2)*. Boston: The MIT Press.

Squazzoni, F. (2012). *Agent-based computational sociology*. Chichester: Wiley.

Terán, O., Rojas, R., Romero, L., & Alarcón, L. (2015b). *Model of the social game associated to the production of potato seeds in a Venezuelan region*. Open ABM. Retrieved on March 2017 from: https://www.openabm.org/model/4606/version/5/view

Terán, O., Sibertin-Blanc, C., & Gaudou, B. (2015a). The influence of moral sensitivity on organizational cooperation. *Kybernetes (Emerald.), 44*(6/7), 1067–1081. Retrieved on March 2017 from: https://doi.org/10.1108/K-01-2015-0015

Velásquez, N. (2001). Desarrollo Sustentable, Modernización Agrícola y Estrategias Campesinas en los Valles Altos Andinos Venezolanos. Los Andes, Escenarios de Cambio a Distintas Escalas. IV Simposio Internacional de Desarrollo Sustentable, November 25 to December 2, School of Computer Science, Mérida, Venezuela.

Part III
Applications in Geography and Urban Development

Chapter 9
Governance of Transitions. A Simulation Experiment on Urban Transportation

Johannes Weyer, Fabian Adelt, and Sebastian Hoffmann

Introduction

The paper at hand presents a simulation framework called SimCo ("Simulation of the Governance of Complex Systems"). SimCo has been used to study the governability of complex socio-technical systems, their dynamics at the macro-level resulting from the interaction of a large number of decision makers at the micro-level. Experiments have been conducted to investigate transitions in network-like infrastructure systems such as urban transportation, which will serve as an example here. In this way various policy measures have been tested, that aim at fostering sustainable transitions.[1]

Concept

SimCo positions itself at the intersection of computer simulation (ABM), governance research (mostly in political sciences) and investigation of infrastructure systems (mostly in the engineering sciences): It refers to common approaches in ABM, stating that the governability of complex socio-technical systems can be investigated experimentally. Adopting insights from governance research, the model differentiates between self-coordination and control. Finally, our framework refers to

---

[1] This paper is an adapted, short version of Adelt et al. (2018).

J. Weyer · F. Adelt (✉) · S. Hoffmann
TU Dortmund University, Faculty of Business and Economics, Technology Studies Group,
Dortmund, Germany
e-mail: johannes.weyer@tu-dortmund.de; fabian.adelt@tu-dortmund.de;
sebastian3.hoffmann@tu-dortmund.de

© Springer Nature Switzerland AG 2019   111
D. Payne et al. (eds.), *Social Simulation for a Digital Society*, Springer
Proceedings in Complexity, https://doi.org/10.1007/978-3-030-30298-6_9

engineering sciences where infrastructures are modelled as specific "spaces" that constrain agents' actions and interactions.

Unlike other simulation frameworks that model specific infrastructure networks, SimCo has been designed as a general-purpose framework that allows to model and to analyse interaction processes in a variety of networked systems. When making their individual choices, agents have to consider the physical infrastructure (with nodes and edges) as an additional variable.

SimCo is rooted in a sociological *macro-micro-macro model* of a socio-technical system, referring to *sociological theories of behaviour and decision-making* (Coleman 1990; Esser 1993, 2000). This model describes system dynamics on the macro level as the emergent result of actors' actions and interactions on the micro level (micro-macro link), which on their part are constrained by the current macro level state of the system (macro-micro link). Taking urban road transportation as an example, users choose routes and modes of transport individually, while being affected by macro constraints like network structure or traffic jam – which is caused by their decisions in previous situations.

Actors' choices are shaped by their individual perception of situational constraints as well as their individual preferences. Hence, choices are influenced by the socio-technical system and the prevailing common preferences as well as the individual actor and his/her specific interests and strategies. Actors make decisions with bounded rationality, referring to multiple evaluation criteria (cf. Velasquez and Hester 2013). Behavioural alternatives are evaluated by assigning a utility to every possible outcome and finally choosing the option with the highest subjective expected utility (cf. Konidari and Mavrakis 2007). This decision-making procedure is conducted by every agent similarly, even if different agent types (see below) assign different utilities to outcomes: every agent calculates all alternatives by summing up the products of his/her individual utility values and the probabilities of reaching these outcomes (which mostly are identical for *all* of the agents).

SimCo has been designed to investigate the risk management of infrastructure systems, aiming at reducing undesirable external effects (e.g. pollution) or avoiding a system breakdown (e.g. congestion), but also the issue of system transformation or regime change (e.g. shifting road transportation towards sustainability).[2]

Research Questions

We have used the SimCo framework to experiment with different scenarios of system transformation in the case of urban road transportation. These scenarios are composed of varying system configurations (e.g. agent populations, technologies,

---

[2] For more details see Adelt et al. (2018) and www.simco.wiwi.tu-dortmund.de.

network topology) and different modes of governance (self-coordination, soft control, strong control). SimCo can help to answer the following questions:

1. Is it possible to construct a stable base scenario of an infrastructure system, rooted in a sociological theory of a macro-micro-macro model and a sociological theory of action that is validated referring to agent and technology types?
2. To which extent do different modes of governance affect the performance of the system and help to achieve policy goals?

Inventory: Social and Technical Components

SimCo has been conceived as an abstract network-like system, that consists of nodes (such as home, place of work, supermarket, train station, charging station etc.), which partly generate payoffs, if an agent reaches them, and edges that connect nodes (such as streets, bus lanes, cycle tracks etc.), which can be used by different technologies (such as cars, bicycles or public transport), but charge the agent with fees. Different social actor groups construct or use these technical components, such as:

- Users, who move through the network in order to fulfil their tasks (such as visiting distinct nodes), using different technologies;
- Operators in the control room, who monitor and manage the system;
- Companies* that offer services, used by agents when moving through the network;
- Politicians* and other stakeholders, who negotiate future policy, decide on the network topology, set up regulations (e.g. emission limits) and provide means for promotion of alternatives;
- Producers* offering technologies, be it established technologies, be it new technologies.

SimCo has been conceived as first step towards modelling the whole system. The present version consists of the network module, the user module and the operator module. It can be complemented by other modules (marked with an asterisk) in future versions.

Every social and every technical component of SimCo has various properties that have been implemented as multi-dimensional variables with mathematical values, which can be freely defined. SimCo thus allows conceiving edges as roads of a transportation network or as transmission lines of an energy grid. The semantics are scenario specific. For our application to urban road transport, we specified money, emissions, capacity and comfort as relevant dimensions. Additionally, these parameters can be used as "levers" of control, e.g. by increasing charges for using streets with old, conventionally fuelled cars. This way, governance measures can be used to restrict the established regime and to promote alternatives.

Interactions

Agents move through the network to fulfil their daily routine tasks, using those technologies, which best meet their individual preferences (e.g. travelling fast, eco-friendly or comfortable). Referring to typical daily routines, we first conceived tasks as "bringing children to the kindergarten", "going to work" and "shopping" on the way back home, but later refrained from using these additional semantics and just assigned three task-nodes to each agent to be visited before returning home. SimCo is not conceived as a reproduction of a real transportation system, but as an abstract representation that refrains from going into every detail.

While travelling through the network, agents interact with nodes and edges and change the state of the network (e.g. by using streets, paying charges or emitting CO_2, cf. Fig. 9.1). Thus, they affect the boundary conditions of other agents, which may be confronted with congested roads or driving bans in polluted areas. Additionally, agents change their own state (e.g. by paying charges, or by receiving income when reaching nodes in the correct order).

As indicated, every parameter can reach certain limits, which triggers interventions such as restrictions or bans for those technologies that are most harmful (referring to the respective parameter).

Apart from their daily routine actions, agents occasionally need to replace a worn-out technology (e.g. buy a new seasonal ticket, a new bicycle or a new car). In our model, 80% opt for that technology they have chosen the most (indicating it fits best their needs), but 20% choose a different technology, which enables us to bring innovative solutions in (cf. Fig. 9.1).

Agents' actions and interactions permanently change the state of the whole system, which evolves dynamically, and thus generate intended or unintended effects – for example regarding political goals such as fostering sustainable mobility.

Governance

The intensity of control can vary heavily. According to our overview of governance research we define governance as "a specific combination of the basic mechanisms of control and coordination in multi-level socio-technical systems" (Weyer et al. 2015: 8). Admittedly, a multitude of definitions exist, ranging from normative to analytical concepts. The *normative*, bottom-up concept uses the term "governance" to demarcate a timely mode of coordination of strategically acting actors, e.g. in policy networks, where the state represents merely one of several co-players (Kooiman et al. 2008; Duit and Galaz 2008; Rhodes 2007). However, since the concept of "policy networks" already has been well established to characterise and analyse this new mode of coordination, we assume that there is no need for another label that refers to the same phenomenon.

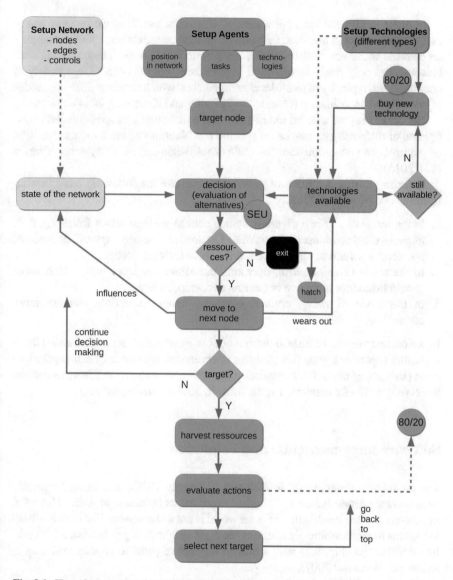

Fig. 9.1 Flow chart representing agents' choices

The *analytical* concept defines governance much more broadly as "the entire spectrum of coexisting modes of collective regulation of societal issues" (Mayntz 2004: 5). Although this neutral approach is capable of identifying a wide variety of governance phenomena, it provokes the question whether labelling a well-known phenomenon with a new term provides any added value, instead of adhering to established concepts such as coordination, actor constellations, and others.

In order to give the term "governance" a specific meaning, we propose not to assign this concept to one specific mode of societal coordination, but to put emphasis on basic social mechanisms such as control or coordination. These mechanisms, however, can only rarely be observed in pure form, but rather in a large variety of combinations, typically in multi-level architectures, which combine different modes of control and coordination (Grande 2012; Kalter and Kroneberg 2014).

Hence we propose a definition of governance that allows an open-minded investigation of different patterns, i.e. of specific combinations of the basic mechanisms of control and coordination in multi-level socio-technical systems (Weyer et al. 2015).

Referring to this definition of governance, we use the following basic mechanisms as components of SimCo:

1. In the mode of *self-coordination*, agents coordinate themselves following their respective rules of decision-making. Controllers monitor system operations, prepared to switch to other modes, if a critical incident occurs.
2. In the mode of *soft control*, operators use stimuli or incentives, which make certain behaviour attractive or unattractive, respectively.
3. In the mode of *strong control*, operators apply constraints that are more compelling.

In the current version of SimCo these modes of governance are implemented as an algorithm (operator agent) that switches interventions off (mode of self-regulation) or on (soft/strong control) automatically, if a relevant parameter reaches pre-defined limits (e.g. 60% of a maximum value for soft, 80% for strong control).

Software Implementation and Validation

The model was programmed in NetLogo (Wilensky 1999), thus using frequently used software for designing sociological experiments by means of ABM. However, in contrast to other models the artificial world is not a chessboard-like lattice, allowing agents to move arbitrarily to one of the eight neighbouring fields, but a network-like structure (cf. Fig. 9.2) with a limited number of paths to choose from and to follow (cf. Wilensky 2007).

To validate SimCo, we repeatedly ran experiments with multiple parameter-variations and analysed them by using different measures, e.g. the payoffs agents gain if reaching task nodes. Furthermore, we tested different network structures, composed of five types of edges that can be used by every mode of transport or by cars, bicycle, or public transport only.

In order to calibrate agent-behaviour, we conducted a survey that provided data on the selection of transport-modes in urban areas, on individual preferences and, finally, on the subjective assessment of respondents, to which degree different modes of transport help to achieve certain goals (e.g. travelling fast or eco-friendly).

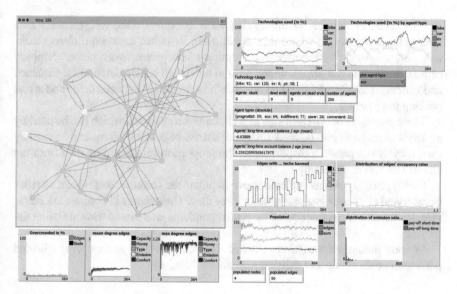

Fig. 9.2 Screenshot (small example network)

Table 9.1 Final basic scenario

	Type	Number/value
Nodes	Home	204
	Task	236
	Standard	160
Edges	Shared-small	984
	Shared-big	110
	Car-only	104
	Bicycle-only	3
	PT-only	110
User-agents	Pragmatic	1000
	Eco	600
	Indifferent	1800
	Penny-pincher	600
	Convenient	2000

We could identify five types of actors: "Pragmatists" mostly favour traveling fast, while "ecos" prefer eco-friendly and affordable transportation. "Penny-pinchers" like travelling cheap and fast, while "convenient" actors favour comfortable transportation. Only "indifferent" agents do not have specific differences in their preferences. By implementing these actor types into agent types we added heterogeneity to our scenario.

The final basic scenario, which runs stable and produces a "standard" picture of present urban transport in a typical mid-sized German town, is summarized in Table 9.1.

This scenario contains various nodes, such as home nodes, where agents start their daily journeys and where they are allowed to switch between different technologies, unspecific task nodes (representing kindergarten, work place, shopping centres or recreational areas), which have to be visited in self-determined sequence and generate payoffs, and standard nodes, which are just in between other nodes and prolong the journey.

There are edges of different size as well, connecting nodes, which can be used by all kinds of technologies (shared-small and shared-big), while others can be used by cars, bicycles or public transport only. The proportions are based on data from the city of Dortmund, Germany.

Five types of agents as depicted above populate the scenario, their characteristics being based on cluster analysis of the survey data. Their shares in contrast are literature driven, as a large fraction of survey respondents was biased towards using the bicycle.

Various parameter variations showed that this specific setting with adjusted parameters runs stable and delivers plausible results.

Experimentation

Experiments included the base scenario (without any intervention) and three governance scenarios with different degrees of intervention. The latter three focus on cars with internal combustion engines and take for granted (without debating in detail) that this technology is more harmful for the environment than other technologies, e.g. considering its CO_2 emissions. These three governance scenarios were implemented as follows:

- Soft control: via road pricing, limited in space and time. Costs of agents using the car are stepwise raised, if traffic jams occur or pollution exceeds a threshold (60% of the maximum value), and lowered again afterwards.
- Strong control: via spatial and temporal bans of cars, if a second threshold (80% of the maximum value) is reached. Agents are forced to change technology or take another route.
- Combination of soft and strong control.

The effects of interventions were measured by means of various indices: mean capacity utilization on edges, mean degree of emission on edges in *short-time* ("days") and *long-time* ("month"), and technology usage (bicycle, car, public transport).

Table 9.2 shows the effects of all three modes of governance on different variables that were used to measure performance. The "mean capacity utilization on edges" measures the percentage of usage (i.e. "occupancy rate") of all edges of the whole network in relation to the respective limit. We distinguish natural limits (e.g. concerning "capacity") and artificial limits (e.g. concerning "emissions"). The former are derived from physical constraints, while the latter could be specified by politicians or the experimenter, for example when trying to reduce the overall

Table 9.2 Results of governance experiments

Governance mode	Mean capacity utilization on edges[a] (%)	Mean emission short-time[a] (%)	Mean emission long-time[a] (%)	Bicycle usage (%)	Car usage (%)	Public transport usage (%)
Self-coordination (basic scenario)	21.36	17.96	33.28	31.61	62.45	5.94
Soft control	15.79	12.76	24.66	46.05	37.48	16.47
Strong control	19.13	15.55	28.92	41.44	52.08	6.47
Soft/strong (combined)	16.37	12.88	24.65	49.94	38.95	11.10

[a]Percentages of respective limits

emissions in a system, "Mean emission" is measured at two time scales, short time (which may be conceived of as one day) and long time (roughly a month). In all three cases, lowering the values of the basic scenario can be regarded as success.

The other three parameters measure the modal shift, i.e. the distribution of car, bicycle and public transport usage. In terms of sustainability, a shift from cars (lowering values) to bicycles and public transport (increasing values) will be regarded as success.

As Table 9.2 shows, we achieve desired effects with all three modes of governance: a decrease of car use and an increase of bicycle and public transport usage, and – triggered by these changes – also a reduction of capacity utilization and emissions. Additionally, soft control mostly performs best (or closely second best). This confirms previous experiments with the simulation framework SUMO-S (Adelt et al. 2014). Only in two cases, the combination of soft and strong control is slightly better. But obviously, political goals of regime change in mobility can best be achieved by relying *only* on soft measures of intervention.[3]

Conclusion

SimCo allows to investigate the issue of governability of complex socio-technical systems experimentally – either referring to the case of risk management or to the case of regime change. We performed experiments on the latter issue and confirmed previous findings, indicating an impact of every mode of control in terms of promoting sustainability. However, again soft control performed better than strong control (or the combination of both modes of governance) in achieving policy goals, the legitimacy of which we simply took for granted.

---

[3] Other experiments might be conducted as well, e.g. relating the effect of different governance modes and various distributions of agent types. Up to now, we didn't conduct this kind of experiments, but can only refer to previous research with a predecessor of SimCo, called SUMO-S. Here we found out indeed, that the success of governance not only depends on the kind of intervention (soft, strong), but also on the mixture of agent types (Adelt et al. 2014).

References

Adelt, F., Weyer, J., & Fink, R. D. (2014). Governance of complex systems. Results of a sociological simulation experiment. *Ergonomics (Special Issue "Beyond Human-Centered Automation"), 57*, 434–448. http://www.tandfonline.com/doi/full/10.1080/00140139.2013.877598.

Adelt, F., et al. (2018). Simulation of the governance of complex systems (SimCo). Basic concepts and initial experiments. *Journal of Artificial Societies and Social Simulation, 21*(2), 2.

Coleman, J. S. (1990). *Foundations of social theory.* Cambridge, MA: Harvard University Press.

Duit, A., & Galaz, V. (2008). Governance and complexity—Emerging issues for governance theory. *Governance, 21*(3), 311–335.

Esser, H. (1993). The rationality of everyday behavior a rational choice reconstruction of the theory of action by Alfred Schütz. *Rationality and Society, 5*(1), 7–31.

Esser, H. (2000). *Soziologie. Spezielle Grundlagen, Bd. 3: Soziales Handeln.* Frankfurt am Main: Campus-Verlag.

Grande, E. (2012). Governance-Forschung in der Governance-Falle?–Eine kritische Bestandsaufnahme. *Politische Vierteljahresschrift, 53*(4), 565–592.

Kalter, Frank/Clemens Kroneberg, 2014: Between mechanism talk and mechanism cult: New emphases in explanatory sociology and empirical research. Kölner Zeitschrift für Soziologie und Sozialpsychologie 66: 91–115.

Konidari, Popi/Dimitrios Mavrakis, 2007: A multi-criteria evaluation method for climate change mitigation policy instruments. Energy Policy 35 (12): 6235–6257. http://www.sciencedirect.com/science/article/pii/S0301421507003229.

Kooiman, J., et al. (2008). Interactive governance and governability: An introduction. *Journal of Transdisciplinary Environmental Studies, 7*, 1–11.

Mayntz, R. (2004). *Governance Theory als fortentwickelte Steuerungstheorie?* (MPIfG Working Paper). http://www.mpifg.de/pu/workpap/wp04-1/wp04-1.html.

Rhodes, R. A. W. (2007). Understanding governance: Ten years on. *Organization Studies, 28*, 1243–1264.

Velasquez, Mark/Patrick T. Hester, 2013: An analysis of multi-criteria decision making methods. International Journal of Operations Research 10 (2): 56–66.

Weyer, J., Adelt, F., & Hoffmann, S. (2015). *Governance of complex systems. A multi-level model (Soziologisches Arbeitspapier 42/2015).* Dortmund: TU Dortmund. http://hdl.handle.net/2003/34132.

Wilensky, U. (1999). *NetLogo.* Evanston, IL: Northwestern University, Center for Connected Learning and Computer-Based Modeling. http://ccl.northwestern.edu/netlogo.

Wilensky, U. (2007). *Link-walking turtles example.* Evanston: Northwestern University, Center for Connected Learning and Computer-Based Modeling. http://modelingcommons.org/browse/one_model/2304.

Chapter 10
Evaluating the Impact of an Integrated Urban Design of Transport Infrastructure and Public Space on Human Behavior and Environmental Quality: A Case Study in Beijing

Liu Yang, Koen H. van Dam, Bani Anvari, and Audrey de Nazelle

Introduction

After the Industrial Revolution, the pattern of cities has changed prominently as a result of urban growth. Urban transport infrastructure (e.g. roads, the railway and light rail) has taken a prominent role in shaping the urban fabric. In developing countries, governments have increasingly invested in infrastructure development to meet the need for rapid urbanization. However, because motorized transportation development has been prioritized over public realm design in many cities, the transport infrastructure sometimes becomes a linear separating component cutting the urban fabric of city centers into fractured, inaccessible and unfriendly pieces of urban spaces. Such spaces, named by Carmona (2003) as Space Leftover After Planning, have a low visual aesthetic quality, minimal public urban facilities, and no meaningful purposes. Therefore, these leftover spaces are less used by citizens for doing physical activity (i.e. recreational activity or active travel), which is a significant element of improving public health. Simultaneously, active travel facilities such as pavements

L. Yang (✉)
Center of Architecture Research and Design, University of Chinese Academy of Sciences, Beijing, China
e-mail: yangliu113@mails.ucas.ac.cn

K. H. van Dam
Department of Chemical Engineering, Imperial College London, London, UK
e-mail: k.van-dam@imperial.ac.uk

B. Anvari
Centre for Transport Studies, University College London, London, UK
e-mail: b.anvari@ucl.ac.uk

A. de Nazelle
Centre for Environmental Policy, Imperial College London, London, UK
e-mail: anazelle@imperial.ac.uk

© Springer Nature Switzerland AG 2019
D. Payne et al. (eds.), *Social Simulation for a Digital Society*, Springer
Proceedings in Complexity, https://doi.org/10.1007/978-3-030-30298-6_10

and bicycle lanes around the motorized transport infrastructures are typically disconnected, leading to inconveniences to walk and cycle in city centers.

Furthermore, the construction of transport infrastructures could cut the urban ecosystem into separate tracts of green spaces, thereby influencing the system's ecological resilience and leaving it vulnerable to shifts in climate change (Alberti and Marzluff 2004). The unconnected blocks in city centers raise an individual's dependence on motorized transportation, resulting in a great amount of air pollution. Bad air quality, in turn impacts on human behavior and results in a reduction of active travel and physical activity in outdoor urban spaces.

It has been widely accepted that an integration of the urban transport infrastructure and land-use is an essential precondition for sustainable development (Varnelis 2008). Some others went a step further demonstrating that transport infrastructures and public spaces should be designed as a holistic system (Ravazzoli and Torricelli 2017). In this respect, this paper focuses on the impacts of integrated transport and public space design on human behavior and the ecosystem. Here, the public space refers to streets, public open spaces (e.g. parks) and public urban facilities (e.g. markets) (UN-Habitat 2015). Initially, this paper aims to explore the relationship between transport-human behavior-ecosystem, leading to a conceptual model. Then we attempt to propose a methodology integrating urban design, computational evaluation of design scenarios, and decision-making support. Moreover, this research aims to show how computer modeling can be used to visualize the impacts of the road network and public space designs on outdoor activities, travel behavior, and car-related air pollution. Finally, the proposed simulation model will be tested on a case study in Beijing in which real-time data validation and air pollution calculation are conducted.

Literature Review

Much research in recent years has focused on the reciprocal relationship between public health and active travel policy, which encourages walking and cycling. In the relation loop, transport and planning policies influence human behavior (in terms of the travel mode choice, route choice and physical activity) and environmental quality, which jointly influence human exposure to the environment (e.g. the exposure to air pollution, heat and traffic injuries) (de Nazelle et al. 2011).

In the domain of urban planning, a group of urban theorists inspired by nature developed a concept of Bio-inspired Urbanism, which regards a city as an evolving organism (Geddes 1915). At present, the self-organization in biological systems has been introduced into the study of urban systems, leading to widespread discussions and practices in exploring the analogy between the built environment or social unit and ecological processes. According to the urban dynamics theory (Forrester 1969), others have treated a city as a complex system consisting of interconnected subsystems—two of which are transport infrastructure and public space systems. On the other hand, Needs Theory (e.g. Mallmann 1980) has been adopted to urban plan in order to achieve a bottom-up planning process. Jackson et al. (2004) argued that a human needs-based construction offers the possibility of sustainable development and "good life" in present-day society.

In terms of the urban design methodology, the traditional esthetic and empirical way of urban design has been criticized for lack of quantitative evaluation. In this respect, mathematical and computational modeling tools have been gradually combined with the traditional linear process of "survey-analysis-plan" to achieve an integrated design-evaluation process, which involves planners, modelers and other designers together in a design process (Batty et al. 2012; Gan 2014). For instance, the transport and land-use sub-systems have been holistically planned and assessed by planners using computer models such as MATSim (Horni et al. 2016) and Integrated Transport-Land Use models (e.g. Zhao et al. 2013). Nevertheless, little work has been done to build up integrated simulation models for transport and public space systems.

One of the initial policy-making models was described by Axelrod (1976), focusing on the causal analysis used by an individual to evaluate sophisticated policy alternatives. The multiplicity of dimensions in a coupled transport-public space system imposes difficulties in determining clear causations by one individual; thus an integrated analysis approach involving scientists, policy-makers, and the public is needed. Moreover, researchers have pointed out that such analysis approach should be designed as an internal part of the decision-making process since stakeholders can judge the feasibility of a method by learning about a system and exploring possible outcomes of different scenarios (Zellner 2008). Since major infrastructure and policy decisions need to be based on sound evidence, computer modeling (with advantages of real-time data validation and prediction) is an efficient way to facilitate the analysis and evaluation of different design scenarios which would then supplement decision-making. Among others, Agent-Based Modeling (ABM) has been markedly applied to provide qualitative insights and quantitative results for the cooperation among decision-makers (Axelrod 1997).

Methodology

In order to explore the impact of transport infrastructure planning on human behavior, and environmental quality, this paper initially analyzes the interactive mechanism underlying transport infrastructure, human behavior and the ecosystem. Needs theories and bio-inspired urbanism theories are adapted to create a conceptual model.

This paper then applies the conceptual model to designing and evaluating an integrated transport-public space scenario, resulting in a methodology for human needs-driven design and computational evaluation. This methodology is in line with the design-evaluation process demonstrated by van Dam et al. (2014) in which an initial design was evaluated using simulation. However, novel steps are made by integrating the traditional linear process of urban design, big data input, and decision-making process. Furthermore, metrics for evaluating the impacts of transport infrastructure and public space designs on human behavior and environmental quality are provided by referring to de Nazelle et al. (2009).

In the part of the computational evaluation, ABM is chosen because it provides a possibility of simulating human needs and behavior by generating a synthetic population from the given statistics, by which the emergence of social behavior could be witnessed. A case study using this computer model is demonstrated afterwards. Aiming at visualizing the traveling of agents between homes, workspaces and public spaces, we adopt and revise the Smart-City Model proposed in (Bustos-Turu et al. 2018; van Dam et al. 2017), which has separate layers of the road network, land-use, and agents.

Initially, a synthetic population as a representative of the entire population of a research field is generated. By using socio-demographic data (the population density, household size, the ratio of worker and non-worker, and car ownership) and geographical data (public spaces, other land-uses, and activity distribution), agents are generated randomly based on the number of people living in each area. Then, activity patterns (i.e. the time-specific sequence of activities throughout a weekday or weekend) are designated for agents which are different for included agent types (e.g. workers, residents who are not economically active, and visitors). In order to categorize human activities according to human needs, five types of activities are elicited: residential activity, industrial activity, commercial activity, cultural activity, and leisure activity. In the land-use layer, the residential activity is given to residential areas, the industrial activity is set to workplaces, and the other three activities are assigned to public spaces. In such, an agent determines when and where to take an activity according to his designated activity patterns and the types of activities provided by each tract of land. In addition, motorized and active travel road networks are provided to agents for traveling to different locations, during which period they follow a shortest path choice algorithm – Dijkstra's algorithm (Skiena 1998) – to choose their routes. Both the equation of activity pattern and the shortest path choice algorithm could be found in the paper (Yang et al. 2019). The model run for a baseline scenario.

To calculate air pollution emitted by vehicles, we track car agents, which are created according to the car-ownership of the total agents and plan their routes only on the motorized road network. Subsequently, emissions on each road are calculated as follows: the traffic volume (the number of agents traveled) on each road segment is recorded hourly, after which the volume is compared with real-time traffic data. By using the software of COPERT, the traffic volume is transformed into the emission rate of NO_2 $\mu g/(s*m)$, which is visualized in an air pollution heat map.

A Conceptual Model

DNA, as a molecule of the self-organized biological system, is similar to the urban system. Inspired by the double helix structure of DNA, a conceptual model of urban infrastructure planning was found (see Fig. 10.1). Corresponding to the two helical

Ecological Needs

Human Needs

──── Transportation System
──── Energy Supply System
──── Water Supply & Drainage System
──── Environmental Sanitation System
──── Disaster Prevention System
──── Telecommunication System

Fig. 10.1 The DNA form of urban infrastructure planning

chains in DNA, human and ecological needs form the backbone of an urban system, since human demands drive the development and functioning of a city, and the ecological quality restricts urban growth. Here, human needs include personal, social, and environmental aspects, and ecological needs refer to biodiversity, networked green and water systems. Depending on the human-ecological needs strands, urban transport infrastructure, as well as the other five types of urban infrastructure systems, should be constructed to meet and connect the two needs, functioning in the same way as the base pairs in DNA. This prototype could support policy-makers by considering the human-ecological needs as vital intervening variables that shape and constrain their actions. Moreover, the needs strands are beneficial for establishing boundaries within which the land-use and transport planning, urban design proposals, public health initiatives, and environmental policies are made.

An Integrated Methodology

To achieve an integrated methodology supporting both a human and ecological needs-driven transport infrastructure design and quantitative evaluation, the traditional design process of "survey–analysis–plan" could be combined with computer simulation. A simulation could be utilized to analyze the mechanism of a given transport infrastructure and public space system aided by big data. Figure 10.2 shows a promising framework, which incorporates human needs in the steps of a survey, system simulation (e.g. individual travel data input), evaluation (by assessing the metrics of physical activity) and decision-making. Ecological needs are incorporated in the steps of problem analysis and evaluation (by determining the parameters of air quality).

In the loop of scenario design, after the physical transport and public space planning, scenarios are set into an agent-based model to simulate their impacts on human behavior (i.e. physical activity pattern, travel mode choice, route choice, and destination choice) and air pollution. Then comparisons with real-time on-road traffic data are carried out in order to validate the model. Moreover, this model could be used to monitor the population's exposure to air pollution, which is a vital metric for evaluating transport scenarios. In this integrated model, public participation appears in the stage of urban context survey, decision-making (i.e. the public is involved by observing the modeling process and providing suggestions) and finally in the long-term

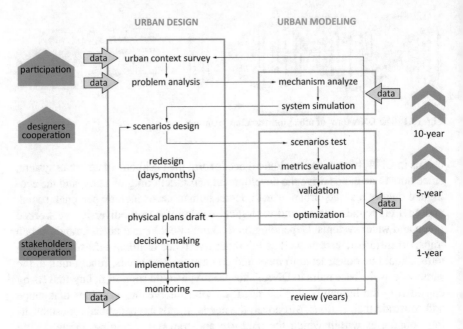

Fig. 10.2 A methodology integrating urban design and modeling to evaluate design scenarios, with giving insights into key performance indicators for policy-making support

monitoring after implementing one of the transport plans. It is worth noting that this method is more than a linear process, but iterative progress that feedback getting from monitoring a designed transport and public space system in 1, 5, or 10 years will yield redesigns of the system.

Case Study and Results

Beijing, the capital of China, is one of the world's most global, dense, and ancient cities. The Beijing-Zhangjiakou (Jing-Zhang) railway, which was the very first urban rail constructed in Beijing, cut through the city center. In order to upgrade the railway for the use of Beijing 2022 Winter Olympic Games, the out-of-date rail track was demolished and will be replaced by a high-speed rail underground. Figure 10.3 shows the research field chosen for this study, a 10-km section of the Jing-Zhang railway, while the status quo of disconnected urban spaces lying around this section is presented in Fig. 10.4. Since the city faces a detrimental problem of heavy air pollution, indicators of environmental impacts and human behavior changes have been gradually taken into account when redesigning transport infrastructures and public spaces in Beijing. To compare various design proposals using metrics such as environmental quality, physical activity, and human exposure could help policy-making. In this paper, we begin by applying an ABM

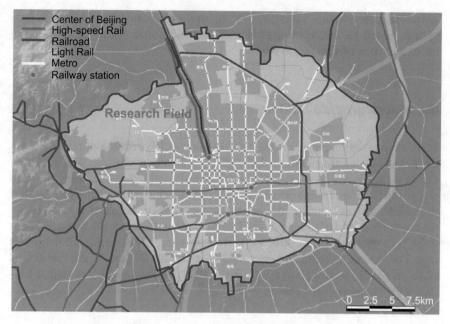

Fig. 10.3 The position of the research field in Beijing

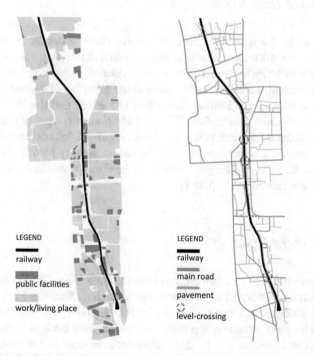

Fig. 10.4 The visual (*left*) and physical (*right*) disconnect of the urban areas around the Jing-Zhang railway

to simulate the baseline scenario of the transport infrastructure and public space system, based on which alternative transport and public space designs can be tested and compared in future work.

Socio-demographic Data Input

The sixth population census of Beijing (Department of Population and Employment Statistics National Bureau of Statistics 2010), which provides demographic distribution at the district level, is used to generate the agent population. Since our case study is located in Haidian district, the number of density (people/ha^2) is assigned to residential areas by taking into account the statistics of permanent residents and the ratio of residential land-use in Haidian district. Furthermore, the synthetic population is categorized by a scale of worker: non-worker: visitor, which is 5:4:1 in this case. The number of households according to household (hh) size in Haidian district (2.46 people/hh) is also considered. Finally, vehicles are generated in each household with a number of 0.6 car/hh.

Geographical Data Input

In order to study the potential effects of various street networks on walking and driving behaviors, the area of an 800 m radius around the railway track with a buffer of 2500–3000 m away from the railway is considered as the case study area. The five types of activities are located on the land-use layer based on two open-source maps: the parcel map for Beijing (Long and Liu 2013) and the POIs (Points of Interest) map of (China-Latest-Free 2017). In addition, two types of road networks are loaded into the model with nearly 7000 road segments. In order to count the total on-road pollutants, an emission standard for cars (Category M*) is given to vehicles, which is 80 g/m NO_2 (Beijing introduced the Euro IV standard and most of the Chinese private cars consume petrol).

Activity Pattern Input

The final input is activity pattern created from the 2008 Time Use Survey (TUS) in China (Department of Social and Science and Technology Statistics National Bureau of Statistics 2008), which was carried out for relevant policy-making and reflecting the Chinese lifestyle. Results in this survey represent seven kinds of human activities on weekdays and weekends, with differences between male and female, urban and rural. Based on this data, an activity schedule is developed for the urban area in Beijing on a typical weekday (see Fig. 10.5) and a weekend (see Fig. 10.6).

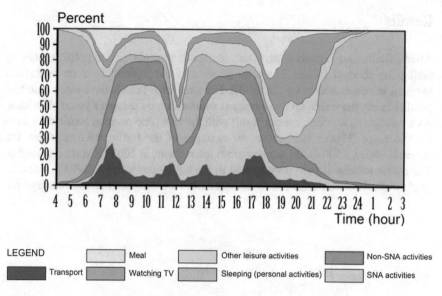

Fig. 10.5 The chart of activity pattern on weekdays. (Adapted from the 2008 Time Use Survey)

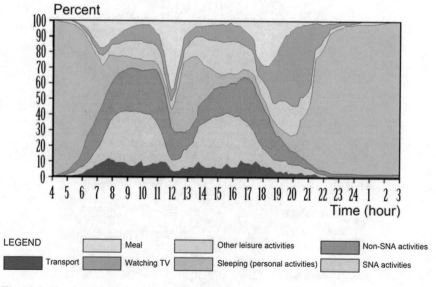

Fig. 10.6 The chart of activity pattern on weekends. (Adapted from the 2008 Time Use Survey)

Results

After initialization, agents depart from homes to workplaces and public spaces by taking the shortest routes. The total number of car agents drove on each road segment is output hourly by the model. For many main roads have more than two parallel lanes, the traffic on these lanes is summed up by using a Python algorithm. As a validation, we compare our result with the real-time on-road traffic data from Google maps. Figure 10.7 shows an example of the traffic volume within the research buffer, 2500–3000 m away the railway, at 10:00 AM on a weekday. The traffic volume is then transformed to emission rates using COPERT software (μg/(s∗m)). Afterwards, the hourly emission rates are assigned to each road segment

Fig. 10.7 Simulation result of the traffic volume within the buffer of 2500–3000 m away from the railway, 10:00 AM on a weekday

Fig. 10.8 Visualization of
the NO$_2$ pollution within
the area of an 800 m radius
around the railway,
10:00 AM on a weekday

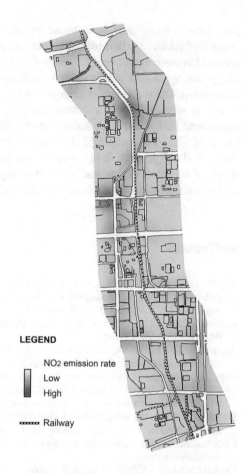

LEGEND

NO2 emission rate
Low
High

▭▭▭ Railway

and visualized in heat maps. Figure 10.8 describes the distribution of NO$_2$ emission
on a neighborhood scale, 800 m away from the railway at the same time as Fig. 10.7.
For the next step, different urban design scenarios could be simulated and compared
by means of changing the input spatial data (e.g. the connectivity of road network
and public spaces distribution).

Discussions and Conclusion

This paper explored the relationship between the transport infrastructure, human
behavior, and ecosystem, resulting in a conceptual model of the urban infrastructure
development inspired by the double helix structure of DNA. Based on this model,
an integrated human needs-driven urban design and computational evaluation methodology with supports for the decision-making process was proposed. Since policy-
making is an interconnected network involving distinct stakeholders, this paper

introduced a transparent agent-based model for the evaluation of the impacts of an integrated urban transport infrastructure and public space design on human behavior and environmental quality. Finally, the evaluation part of the integrated methodology was applied in a case study in Beijing to simulate the effects of a baseline transport-public space design scenario on local air quality.

Future work will focus on testing different design scenarios and evaluating their impacts on human behavior and air quality. This could eventually lead to a comparison of different integrated transport-public space plans for supporting the policy-making process. It is expected that our methods can be applied to a wide range of transport design and evaluation applications to satisfy both human and ecological needs and to involve civic participatory in the policy-making process.

References

Alberti, M., & Marzluff, J. M. (2004). Ecological resilience in urban ecosystems: Linking urban patterns to human and ecological functions. *Urban Ecosystem, 7*(3), 241–265. https://doi.org/10.1023/B:UECO.0000044038.90173.c6.

Axelrod, R. M. (1976). *Structure of decision: The cognitive maps of political elites*. Princeton: Princeton University Press.

Axelrod, R. M. (1997). *The complexity of cooperation: Agent-based models of competition and collaboration*. Princeton: Princeton University Press.

Batty, M., Axhausen, K. W., Giannotti, F., Pozdnoukhov, A., Bazzani, A., Wachowicz, M., et al. (2012). Smart cities of the future. [Article]. *European Physical Journal-Special Topics, 214*(1), 481–518. https://doi.org/10.1140/epjst/e2012-01703-3.

Bustos-Turu, G. (2018). Integrated modelling framework for the analysis of demand side management strategies in urban energy systems, PhD Thesis, Imperial College London, October 2018

Carmona, M. (2003). *Public places, urban spaces: The dimensions of urban design*. Oxford: Architectural Press.

China-Latest-Free. (2017). *Geofabrik GmbH*. https://download.geofabrik.de/asia/china.html. Accessed 28 Feb 2018.

de Nazelle, A., Nieuwenhuijsen, M. J., Antó, J. M., Brauer, M., Briggs, D., Braun-Fahrlander, C., et al. (2011). Improving health through policies that promote active travel: A review of evidence to support integrated health impact assessment. *Environment International, 37*(4), 766–777.

de Nazelle, A., Rodríguez, D. A., & Crawford-Brown, D. (2009). The built environment and health: Impacts of pedestrian-friendly designs on air pollution exposure. *Science of the Total Environment, 407*(8), 2525–2535.

Department of Population and Employment Statistics National Bureau of Statistics. (2010). *Tabulation on the 2010 population census of the People's Republic of China*. Beijing, China: China Statistics Press and Beijing Info Press.

Department of Social and Science and Technology Statistics National Bureau of Statistics. (2008). *2008 time use survey in China*. Beijing, China: China Statistics Press.

Forrester, J. W. (1969). *Urban dynamics*. Cambridge, MA: MIT Press.

Gan, W. (2014). *Responsive urban simulation: An approach towards real time evaluation of urban design projects* (Master's thesis). Politecnico Di Milano, Milano, Italy.

Geddes, S. P. (1915). *Cities in evolution: An introduction to the town planning movement and to the study of civics*. London: Williams & Norgate London.

Horni, A., Nagel, K., & Axhausen, K. W. (2016). *The multi-agent transport simulation MATSim*. London: Ubiquity Press.

Jackson, T., Jager, W., & Stagl, S. (2004). Beyond insatiability: Needs theory, consumption and sustainability. *ESRC Sustainable Technologies Programme Working Paper Series*, 2.

Long, Y., & Liu, X. (2013). *Automated identification and characterization of parcels (AICP) with OpenStreetMap and points of interest* (Working paper # 16). Beijing City Lab.

Mallmann, C. (1980). Society, needs and rights: a systemic approach. In K. Lederer & J. V. Galtung (Eds.), *Human needs: A contribution to the current debate* (pp. 37–54). Cambridge, MA: Oelgeschlager, Gunn and Hain.

Ravazzoli, E., & Torricelli, G. P. (2017). Urban mobility and public space. A challenge for the sustainable liveable city of the future. *The Journal of Public Space, 2*(2), 37–50.

Skiena, S. S. (1998). *The algorithm design manual*. London: Springer Science & Business Media.

UN-Habitat. (2015). *Global public space toolkit: From global principles to local policies and practice*. Nairobi: United Nations Human Settlements Programme.

van Dam, K. H., Bustos-Turu, G., & Shah, N. (2017). A methodology for simulating synthetic populations for the analysis of socio-technical infrastructures. In W. Jager et al. (Eds.), *Advances in social simulation 2015* (Vol. 528). Cham: Springer.

van Dam, K. H., Koering, D., Bustos-Turu, G., & Jones, H. (2014). *Agent-based simulation as an urban design tool: Iterative evaluation of a smart city masterplan*. In The Fifth Annual Digital Economy All Hands Conference.

Varnelis, K. (2008). *The infrastructural city: Networked ecologies in Los Angeles*. Barcelona: Actar.

Yang, L., Zhang, L., Stettler, M. E. J., Sukitpaneenit, M., Xiao, D., & van Dam, K. H. (2019). Supporting an integrated transportation infrastructure and public space design: A coupling simulation methodology for evaluating traffic pollution and microclimate. Sustainable Cities and Society. https://doi.org/10.1016/j.scs.2019.101796

Zellner, M. L. (2008). Embracing complexity and uncertainty: The potential of agent-based modeling for environmental planning and policy. *Planning Theory & Practice, 9*(4), 437–457. https://doi.org/10.1080/14649350802481470.

Zhao, P., Chapman, R., Randal, E., & Howden-Chapman, P. (2013). Understanding resilient urban futures: A systemic modelling approach. *Sustainability, 5*(7), 3202–3223. https://doi.org/10.3390/su5073202.

Chapter 11
Prescription for Urban Sprawl. Street Activeness Changes the City

Hideyuki Nagai and Setsuya Kurahashi

Introduction

Throughout the twentieth century, there was a rapid increase in the world's population along with rapid urbanization. Even in this century, there is no decline in the momentum (United Nations 2014). Urban sprawl structure has been one of the large themes related to urbanization for decades and recognized as a serious issue in many cities (Haase et al. 2010; Kazepov 2011). In Japan, from the beginning of the twentieth century to the post-war's high economic growth period, many bedroom towns were developed on the periphery of large cities. The residents in those towns were assumed to commute by public transportation, but gradually motorization in their daily life progressed. For this reason, the expansion of low-density urban areas has been progressing until now (Kaido 2005; Millward 2006). There is a concern that this situation may lead to consequences such as a decline in living convenience of residents due to a lack of public services. Additionally, another concern is problems caused by the excessive dependence on automobiles. This results in another chain reaction as air pollution and increase of consumption of fossil fuels. There is a concern that in the near future such problems will become more serious in many cities of not only Japan but also some emerging countries. Therefore, as countermeasure to this situation, transformation into compact cities has been explored (Howley et al. 2009). From the point of view of regarding a city as the dynamism where autonomous agents such as individuals, families and companies (Howley et al. 2009), however, the difficulty is highlighted in direct control of it. With this in

H. Nagai (✉)
Research Group on Fusion of Informatics and Social Science, University of Tsukuba,
Tokyo, Japan
e-mail: haabiz@ae.auone-net.jp

S. Kurahashi
Graduate School of System Management, University of Tsukuba, Tokyo, Japan
e-mail: kurahashi.setsuya.gf@u.tsukuba.ac.jp

© Springer Nature Switzerland AG 2019   135
D. Payne et al. (eds.), *Social Simulation for a Digital Society*, Springer
Proceedings in Complexity, https://doi.org/10.1007/978-3-030-30298-6_11

mind, in this study we conducted an agent-based model (ABM) to verify the possibility of changing an urban structure toward a desirable formation indirectly, by inducing behaviors of autonomous individual residents rather than by forcing them. In ABM, interactions between individual agents and the environment are reflected. For this reason, this simulation method is expected to contribute to measuring the effect of policies on a complex environment (Jager and Mosler 2007; Taniguchi et al. 2003).

Related Studies

Changes of Urban Structure Through Residents' Behavior

The necessity to coordinate policies to compactify an urban structure and mobility management measures (Taniguchi et al. 2003), the possibility that an urban structure would naturally transform into a compact one, through mobility management measures to decrease automobile-dependent trend (Fujii and Someya 2007) and the land use model of each household's residential relocation through considering land rent and distance from residence to work place (Togawa et al. 2008) were raised. Based on the above studies, Taniguchi and Takahashi (2012) verified the fact that soft measures that control automobile use changed an urban structure through travel mode selection and residential relocation by individuals, by using ABM. This suggested the possibility that an urban structure can be brought about indirectly toward a desirable formation by interfering with the daily travel of residents. For this reason, they contributed to a long-term prediction of effect of policies on urban issues.

On the other hand, in recent years, the importance of informal public spaces that was once proposed in such as Jacobs (1961) has also been revaluated (Zukin 2009). With this in mind, in this study we considered the existence of informal public spaces other than residences and work places, such as facilities, where everyone feels free to stop off and streets, and the qualitative benefits that can be gained by staying in such places. These facilities indicate, for example, Takeo City Library in Saga, Japan (2013), Gifu Media Cosmos in Gifu, Japan (2015), Idea Stores in London, UK (2002–), etc.

Position of this Study

With these in mind, in this study we made an ABM that assumed travel where resident agents stop off at an informal public facility while commuting on a simple urban model. With this model, we conducted experiments to verify the possibility as to whether changes in the location of such a facility, along with a promotion of street activeness around such a facility, could indirectly change an urban structure toward a desirable formation through daily travel and residential relocation by residents. Additionally, in the same way we conducted combined experiments to verify another case when implementing the policy on transportation to improve the urban environment.

Simulation Model

Urban Model

Figure 11.1 shows the schematic of the simple urban model. This is the simplified expression of a part of a central business district and bedroom towns connected by railway. In the urban model, two domains are located: the residence zone and the destination zone. The residence zone is an aggregation of residences, which are the base point of each resident agent's travel corresponding to their daily commute. The destination zone is an aggregation of work places or schools, which are also a half-way point of the travel. The distance between the two centers of each zone is 4 km at the same latitude. Two train stations are located at the center of each zone, and these stations are connected by railway. With the assumption that uniform and high-density sidewalks and roads are located on this continuous planar space, resident agents can freely move on this space on foot, by bicycle or car. As the initial location, in the residence zone, residences of the same number as the number of resident agents are randomly located based on normal distribution centering on the stations. One resident agent corresponds to ten households in the real world. Similarly, destinations of the same number are also located centering on another station.

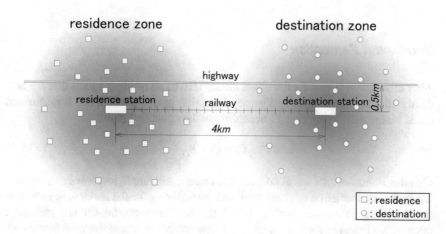

Fig. 11.1 Simple urban model

Daily Travel

The series of travels of each resident agent from the base point to the final destination is referred to as a linked trip. Each travel mode that is a component of a linked trip is referred to as an unlinked trip. Additionally, a main travel mode among unlinked trips of a certain linked trip is referred to as a representative travel mode of the linked trip. After that, each resident agent repeats travel assumed as a commute according to the selected linked trip every day. The initial representative travel mode of all resident agent is train. Departure time of each resident agent, from its residence and destination, is based on normal distribution. At the time when each resident agent comes back residence, total travel cost C_i is calculated according to the equation below.

$$C_i = w_t C_t + w_c C_c + w_f C_f \qquad (11.1)$$

C_t, C_c and C_f indicate time cost, charge cost and fatigue cost. Similarly, w_t, w_c and w_f indicate each preference bias. The preference biases of all agents are assumed to be equal. According to the cost, the resident agent changes the values V_i of i-th linked trip, according to the equation below.

$$V_i \leftarrow a(-C_i) + (1-a)V_i \qquad (11.2)$$

The following day's travel mode of the resident agent is selected by the epsilon-greedy method based on this value. e is gradually attenuated according to the equation below. And each resident agent fixes the travel mode in one way through a learning period of 30 days.

$$e \leftarrow re \qquad (11.3)$$

Residential Relocation

After all resident agents fix the travel mode, 1/10 of all resident agents that are randomly chosen change their residences. Selection of a new residence is performed based on the total living cost $_iC_i$ according to the equation below.

$$_iC_i = C_i + R_i \qquad (11.4)$$

C_i and R_i indicates total travel cost and land rent. Land rent R_i increases corresponding to the accumulation of residences and destinations. To the resident agents chosen to change their residences, 10 of the residence candidates are presented randomly. Those resident agent selects a new residence where the total living cost $_iC_i$ is the minimum. After 20 times of the loop process of residential relocation, model stops running.

Conducting Experiments

The result of each scenario was observed under the following indicators and estimated about change of urban structure.

- Percentage of each representative travel mode
- Total CO_2 emission (expressed as percentage relative to the basic model)
- Average travel hours
- Standard deviation of distribution of residences (x and y coordinate)
- Distribution map of residences

Setting values of parameters of the model were set with statistical data of the Ministry of Land, Infrastructure and Transport and public transportation services. In this study, we advance experiments and discussion based on the values of parameters, whereas different values are enabled to set depending on the other urban area.

Experiment 1 – Basic Model

Hereinafter, the experiment using the basic model is referred to as scenario A. Table 11.1 shows the quantitative initial state and result of scenario A. Figure 11.2 shows the final distribution of residences of scenario A. The result shows that the percentage of car users reached over 85%, the majority of whose residences were distributed on the periphery of the destination zone.

The result of the experiment using the basic model shows that an urban structure changed significantly, from an initial zoned structure between residence zone and destination (work place) zone, to a sprawl structure of residences on the periphery of the destination zone. Additionally, the majority of resident agents who were living in such residences used car as a main travel mode (pink-colored dots). This almost coincides with the experimental results of Taniguchi and Takahashi (2012). Additionally, this coincides with the fact that the main commuting mode has shifted from railway-centered to private car-centered in many local cities in Japan. These were reproduction of multiple social patterns that were not built in the model. Therefore, these show the validity of our basic model (Railsback and Grimm 2011).

Table 11.1 Result of experiment 1

| Scenario | Percentage of representative travel modes | | | | CO_2 emission | Travel hours | Standard deviation | |
	Walk (%)	Bicycle (%)	Train (%)	Car (%)	(%)	(min)	x-cor	y-cor
Initial state	0.0	0.0	100.0	0.0	19.8	47.8	8.0	8.0
A	1.3	2.5	9.4	86.8	100.0	10.4	23.1	9.7

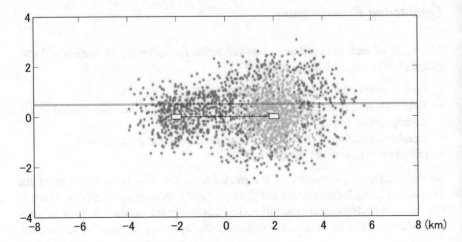

Fig. 11.2 Final distribution of residences of scenario A

Experiment 2 – A Facility for Stopping Off and Street Activeness

In these experiments, one public facility, such as the libraries mentioned before, is located inside or outside the destination zone. And from each destination, all resident agents leave for not each residence but the facility. After arriving and staying there, finally they leave for each residence. Staying time is based on a normal distribution. Within 500 m radius centering on the facility, which was set as an influential area of the facility, a promotion of street activeness is considered. Locations of the facility are assumed to be the following three types as shown in Fig. 11.3.

- B: 2 km south and 0.5 km east from the destination station
- C: same place as the destination station
- D: 0.5 km south and 0.5 km east from the destination station

When each resident agent moving on foot or by bicycle moves within 500 m radius centering on the facility for stopping off which was set as an influential area of the facility, according to the number of other resident agents moving on foot or by bicycle moves within 100 m radius centering on the relevant resident agent, D_{st} (agent), street activeness, P (yen/min), is calculated as an incentive to such a pedestrian or cyclist, according to the equation below.

$$P = \min\left(h_{st}D_{st}, P_{\max}\right) \tag{11.5}$$

The total travel cost is reduced by the amount obtained after multiplying the P with preference bias w_P according to the equation below.

$$C_i = w_t C_t + w_c C_c + w_f C_f - w_P P \tag{11.6}$$

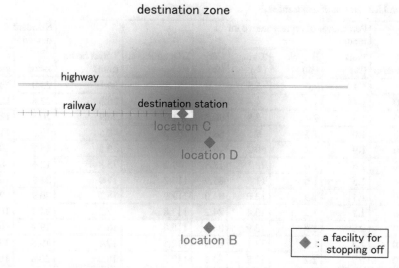

Fig. 11.3 Types of the facility location

The coefficient of street activeness, h_{st}, can be regarded as a scale of efforts to produce further street activeness within the influential area of the facility for stopping off, according to accumulation of pedestrians. This coefficient is enhanced by arranging comfortable sidewalks and cycle roads, attractive retail stores or holding attractive events. When this coefficient is raised, the advantage of moving on foot or by bicycle within the relevant area increases; therefore, the amount of the reduced total travel cost increases. Therefore, this coefficient can be regarded as a coefficient of gain.

These experiments were conducted under the conditions using a combination of the above mentioned three ways for facility location, and the six ways for coefficient of street activeness, 0, 10, 20, 30, 40 and 50. Hereinafter, each of these experiments is expressed, for example scenario C10, by symbols of B through D, indicating the location of the facility for stopping off, and the coefficient of street activeness.

Results of Experiment 2

Table 11.2 shows the quantitative result of scenario B0 to 50, C0 to 50 and D0 to 50. The results of scenario B0, C0 and D0, location of a facility for stopping off was changed, while promotion of street activeness was not implemented, show that the percentage of car users reached over 85%, the majority of whose residences were distributed as a sprawl structure like the result of scenario A.

The results of scenario B0 to 50 show that change was almost not observed in B0 to 30. However, when scenario reached B40 with further advancement of the

Table 11.2 Result of experiment 2

Scenario	Percentage of representative travel modes				CO_2 emission (%)	Travel hours (min)	Standard deviation	
	Walk (%)	Bicycle (%)	Train (%)	Car (%)			x-cor	y-cor
B0	1.1	2.0	5.3	91.6	151.2	36.1	23.0	13.4
B10	1.4	1.7	4.8	92.1	153.2	36.3	22.4	12.9
B20	1.5	2.4	5.8	90.3	151.6	36.9	20.8	12.1
B30	1.0	1.5	9.5	88.0	152.2	40.5	19.5	11.8
B40	1.9	1.2	50.2	46.7	87.5	56.5	15.9	10.2
B50	2.6	1.0	78.1	18.3	49.2	75.8	12.6	8.8
C0	1.5	1.6	11.0	85.9	115.2	17.9	21.9	10.5
C10	1.1	1.9	13.0	84.0	112.9	18.2	20.9	9.8
C20	1.2	1.2	13.4	84.2	113.8	17.2	21.4	10.3
C30	0.9	1.4	12.9	84.8	116.6	18.3	19.9	10.8
C40	1.3	2.7	13.1	82.9	112.5	17.1	19.8	10.5
C50	0.9	2.2	13.8	83.1	117.5	18.0	20.8	10.2
D0	0.9	1.4	8.5	89.2	126.5	22.9	23.5	11.1
D10	1.3	1.3	8.0	89.4	130.8	24.2	23.6	10.9
D20	1.3	2.0	13.6	83.1	127.6	25.3	19.8	10.1
D30	1.3	1.6	42.1	55.0	94.7	29.8	15.3	8.6
D40	1.8	1.2	69.0	28.0	58.4	41.2	13.4	8.5
D50	2.8	0.8	82.8	13.6	39.6	49.2	11.4	8.5

promotion of street activeness, about a half of the car users changed their travel mode to using train. With this, about the distribution of residences, the remarkable phase transition emerged, that the sprawl structure was improved on the periphery of the destination zone and the cluster of residences of train users was formed around the residence station. With this phase transition, the total CO_2 emission was reduced considerably. Furthermore, with further advancement of the promotion of street activeness, this trend advanced.

The results of scenario C0 to 50 show that car users' change to other travel modes and change of the distribution of residences were not observed although scenario reached C50 with further advancement of the promotion of street activeness.

The results of scenario D0 to 50 show that large number of car users changed their travel mode to using the train or walking when scenario reached D30, and the majority of car users also changed their travel mode when scenario reached D50. With this, in scenario D50, as shown on Fig. 11.4, the cluster of residences of train users (green-colored dots) around the residence station became more remarkable and the total CO_2 emission was reduced to about 40% of scenario A.

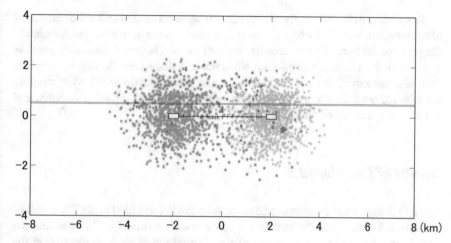

Fig. 11.4 Final distribution of residences of scenario D50

Discussion of Experiment 2

In any cases except when introducing a facility for stopping off at the same place as the destination station, at the point when the implementation of the promotion of street activeness around the facility reached a certain scale, the urban structures were improved drastically. This suggests that such a promotion is not effective as far as not implemented on a certain large scale. And among such cases, in the cases when introducing the facility near the station, an earlier improvement of urban structure was observed. In contrast, in the cases when introducing the facility at the same place as the station, the improvement was not observed. These results suggested the possibility that the slight difference in location of the facility might bring about the significant difference in the future urban structure.

Experiment 3 – An Urban Sprawl Structure as an Initial State

The discussion in the previous section focused on maintaining the given zoned compact urban area. However, especially in Japan, the large part of the land is mountainous, thus land suitable for urbanization is relatively small (Koike et al. 2000). Additionally, the population is declining. For these reasons, it may be unrealistic to assume that we create a compact city newly and maintain it. Instead, improvement of many cities that have already sprawled is considered to be more important. With this in mind, in this section, we consider the final state of scenario A, as shown on Fig. 11.2, as a sprawl structure of residences where residents use a private car as a main travel mode. Then we conduct the experiments where this state is an initial state, and on that basis, the policies similar to those are implemented in the previous section.

To simplify a discussion, these experiments were conducted under the conditions of a combination of the facility location D, which was the most effective location in the previous section, and the same six ways for the coefficient of street activeness as experiment 2, in addition to the case when not introducing the facility. Hereinafter, each of these experiments is expressed, for example, scenario SD20, by combining the initial letter S for the word sprawling, symbols of A and D used previously, and the coefficient of street activeness.

Results of Experiment 3

Table 11.3 shows the quantitative result of scenario SA and SD0 to 50. The result of scenario SA shows that the percentage of car users increased further and reached over 90%, and the sprawl structure of the distribution of their residences on the periphery of the destination zone advanced. The results of scenario SD0 to 50, which are not like the result of the series of D in the previous section, show that car users' change to other travel modes and change of the distribution of residences were not observed although the promotion of street activeness was advanced. As a result, like scenario SA, the percentage of car users increased further, and the sprawl structure of the distribution of their residences advanced, as shown on Fig. 11.5.

Discussion of Experiment 3

In the cases when setting the sprawl structure of residences where residents commuted by their private car as an initial state, progressing of sprawling of urban structure could not be stopped, even with the combinations of introducing the facility for stopping off and implementing the promotion of street activeness around the

Table 11.3 Result of experiment 3

Scenario	Percentage of representative travel modes				CO_2 emission (%)	Travel hours (min)	Standard deviation	
	Walk (%)	Bicycle (%)	Train (%)	Car (%)			x-cor	y-cor
SA	1.5	2.7	3.0	92.8	92.3	6.8	26.8	11.0
SD0	1.3	1.3	1.8	95.6	116.3	18.7	25.9	18.7
SD10	1.5	2.0	3.1	93.4	113.9	18.7	25.2	12.7
SD20	1.5	2.7	3.1	92.7	111.1	19.4	25.1	12.8
SD30	1.6	1.9	3.0	93.5	114.3	19.9	25.5	12.6
SD40	0.9	1.0	3.1	95.0	116.8	19.3	25.6	13.0
SD50	1.7	2.7	3.5	92.1	114.3	20.1	25.9	12.5

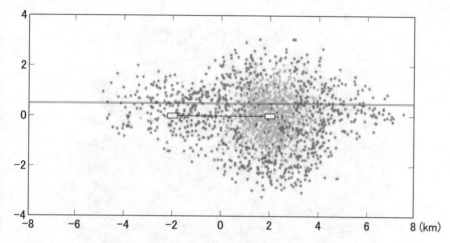

Fig. 11.5 Final distributions of residences of scenario SD50

facility. This suggests that once residents established the lifestyle that they commute by their private car and live near their destination, the urban sprawl structure becomes strong and very difficult to be upset.

Experiment 4 – Introduction of Tramway

Outline of Experiment 4

One of the policies to reduce traffic congestion, save energy and reduce air contamination in an urban area is the introduction of a tramway. A tramway is mainly used for short distance trip and is also characterized by its ease of introduction compared to normal railway. For these reasons, they have been introduced in many European cities (Aleta et al. 2017). However, in recent years, momentum for the introduction of tramways has also risen in Japan. With this in mind, in this section we consider such tramways to the residences. These three tramway lines are laid centering on the destination station as shown in Fig. 11.6. And tramway stations are located at 400 m intervals along each line. Along with this, residents can also choose two types of linked trips: one is the trip where a resident commutes by tram (and on foot or by bicycle), the other is the trip where a resident commutes by train and tram in combination.

These experiments were conducted under the conditions of a combination of the facility location D as in the previous section and the same six ways for the coefficient of street activeness as in experiments 2 and 3, in addition to the case when not introducing the facility for stopping off. Hereinafter, each of these experiments is expressed, for example, scenario SD20t, by combining the initial letter S for the word sprawling, previously-used symbols of A and D, the coefficient of street

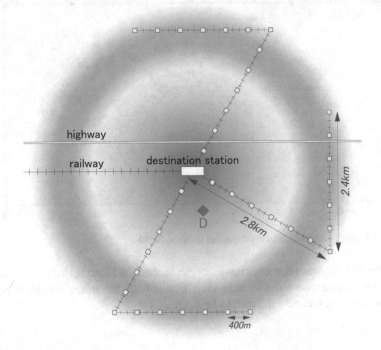

Fig. 11.6 Schematic of tramway lines

activeness and the initial letter t for the word tramway. Additionally, scenario SD50t+, which was run for the double period of SD50t, was executed.

Results of Experiment 4

Table 11.4 shows the quantitative result of scenario SAt, SDt0 to 50 and SDt50+. The result of scenario SAt, when compared with the result of scenario SA in the previous section, shows that car users decreased by about 10 points, while tram users increased accordingly. However, like scenario A, the sprawl structure of the distribution of their residences on the periphery of the destination zone also advanced.

The results of scenario SD0 to 50t, when compared with the results of the series of SD, show that car users decreased by about 10 points in SD0t, while tram users increased accordingly. Additionally, in advancing the promotion of street activeness, from scenario SD30t, car users started decreasing. And when experiment reached SD50t, the percentage of tram users reached over 70% in total.

Furthermore, the results of scenario SD50t+, where scenario SD50t was run further, shows that the percentage of tram users reached over 90% in total. With this,

Table 11.4 Result of experiment 4

	Percentage of representative travel modes						CO_2 emission	Travel hours (min)	Standard deviation	
Scenario	Walk (%)	Bicycle (%)	Train (%)	Car (%)	Train + train	Tram			x-cor	y-cor
SAt	1.6	3.7	5.0	81.6	3.3	4.8	83.9	10.6	27.1	11.2
SD0t	1.2	4.7	3.4	84.0	3.6	3.1	122.9	22.1	26.2	12.6
SD10t	1.3	2.2	5.5	82.0	4.8	4.2	120.0	23.7	26.8	12.3
SD20t	1.1	2.2	5.3	73.3	8.0	10.1	108.3	25.8	26.1	12.2
SD30t	1.6	1.4	4.1	49.0	18.3	25.6	79.1	33.1	25.7	12.5
SD40t	1.4	1.9	5.3	29.6	24.2	37.6	57.1	43.5	25.5	12.5
SD50t	1.2	1.3	5.2	19.0	31.3	42.0	43.3	51.2	24.9	12.2
SD50t+	1.9	0.6	1.7	1.4	30.5	63.9	17.4	53.9	23.7	14.0

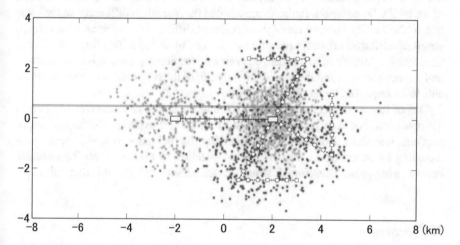

Fig. 11.7 Final distributions of residences of scenario SD50t+

the total CO2 emission was also reduced considerably and the distribution of residences was divided into following two, as shown on Fig. 11.7: One is the cluster of residences, where residents commute by train and tram in combination, on centering the residence station (orange-colored dots). The other is the wide distribution of residences, where residents commute by tram, from the inside to the periphery of the destination zone along tramway lines (brown-colored dots).

Discussion of Experiment 4

In the cases of tramway lines in combination with introducing the facility for stopping off and promotion of street activeness, almost all car users shifted their travel mode to tram. Along with this, although the sprawl structure of residences was not improved, traffic congestion and air contamination in the urban area were reduced. Therefore, as for this distribution of residences from the inside to the periphery of the urban central area, this can be also evaluated positively as the achievement of living in the urban central area, where residents live and work conveniently and additionally they are also expected to activate the city.

Conclusions

We proposed an agent-based model for urban dynamics combining introducing of a facility for stopping off and promotion of street activeness around the facility on an appropriate scale. The model clarified that measures could affect in compactification of an urban structure physically and decrease of car users along with reduction of the total CO_2 emission. It also demonstrated that the slight difference in the location of the facility brought about the significant difference in such an effect. Next, we clarified that an urban sprawl structure formed by taking a long time was strong, therefore it was difficult to be changed even with the above policies. We also showed that the environment and the quality of daily life of the residents, however, might be able to be improved by combining tramway.

One of future perspectives is an extension of the model to discuss about changes in population. For example, in this study, in each of the experiences, the number of residents was fixed. However, when observing the real cities as a wide area, overcrowding by excess of moving in and depopulation by excess of moving out have become a big issue. Therefore, it would be desirable to consider this dynamism.

References

Aleta, A., Meloni, S., & Moreno, Y. (2017). A multilayer perspective for the analysis of urban transportation systems. *Scientific Reports, 7*, 44359.

Fujii, S., & Someya, Y. (2007). A behavioral analysis on relationship between individuals' travel behavior and residential choice behavior. *Infrastructure Planning Review, Japan Society of Civil Engineers, 24*, 481–487.

Haase, D., Lautenbach, S., & Seppelt, R. (2010). Modeling and simulating residential mobility in a shrinking city using an agent-based approach. *Environmental Modelling & Software, 25*(10), 1225–1240.

Howley, P., Scott, M., & Redmond, D. (2009). An examination of residential preferences for less sustainable housing: Exploring future mobility among Dublin central city residents. *Cities, 26*(1), 1–8.

Jacobs, J. (1961). *The death and life of great American cities*. New York: Random House.

Jager, W., & Mosler, H. J. (2007). Simulating human behavior for understanding and managing environmental resource use. *Journal of Social Issues, 63*(1), 97–116.

Kaido, K. (2005). Urban densities, quality of life and local facility accessibility in principal Japanese cities. In M. Jenks & N. Dempsey (Eds.), *Future forms and design for sustainable cities*. Amsterdam/Boston: Architectural Press.

Kazepov, Y. (2011). *Cities of Europe: Changing contexts, local arrangement and the challenge to urban cohesion* (Vol. 46). Somerset: Wiley.

Koike, H., Morimoto, A., & Itoh, K. (2000). A study on measures to promote bicycle usage in Japan. In *VeloMondial 2000-world bicycle conference* (pp. 19–22).

Millward, H. (2006). Urban containment strategies: A case-study appraisal of plans and policies in Japanese, British, and Canadian cities. *Land Use Policy, 23*(4), 473–485.

Railsback, S. F., & Grimm, V. (2011). *Agent-based and individual-based modeling: A practical introduction*. Princeton, Princeton University Press.

Taniguchi, A., Shin'ei, T., & Hara, F. (2003). Attempt of travel feedback programs aimed at smart car using. *Communications of the Operations Research Society of Japan, 48*(11), 814–820.

Taniguchi, T., & Takahashi, Y. (2012). Multi-agent simulation about urban dynamics based on a hypothetical relationship between individuals' travel behavior and residential choice behavior. *Transactions of the Society of Instrument and Control Engineers, 47*, 571–580.

Togawa, T., Hayashi, Y., & Kato, H. (2008). An expansion of equilibrium type land use model by multi agent approach. In *37th the Committee of Infrastructure Planning and Management (IP)*, Japan Society of Civil Engineers.

United Nations. (2014). *World urbanization prospects: The 2014 revision*. http://esa.un.org/unpd/wup/Highlights/WUP2014-Highlights.pdf. Archived at: http://www.webcitation.org/6hhLAUhZx

Zukin, S. (2009). *Naked city: The death and life of authentic urban places*. New York: Oxford University Press.

Chapter 12
The Greater Dublin Region, Ireland: Experiences in Applying Urban Modelling in Regional Planning and Engaging Between Scientists and Stakeholders

Laura O. Petrov, Brendan Williams, and Harutyun Shahumyan

Introduction

The role of the university is recognized as a key partner to support with knowledge and tools the creation of a better understanding of the challenges facing regions and cities internationally (Hurley et al. 2016). Within applied research it is evident that the development of stronger links between scientists and stakeholders is essential. In many regions, models, scenarios, and indicators are used as supporting tools for urban and regional planning and development. These urban modelling approaches are also used for related policies and agreements/negotiations. However, many times the scientists neglect to provide feed-back as to how they used the stakeholder's inputs and what were the results and vice-versa.

The case study used for this work is the Greater Dublin Region (GDR), Ireland which was the principal beneficiary of the 'Celtic Tiger' period of growth in Ireland (Couch et al. 2007; Gkartzios and Scott 2005). As a consequence this region experienced the greatest transformations through the period from the 1990s with economic and population growth in the Dublin area leading to a rapid urban growth, particularly in South Dublin (9.6%) and Fingal (21.1%), to the north (Williams et al. 2010). This was influenced by several push factors such as high house prices and an inadequate transport system combined with weak planning which rapidly drove urban development outwards into the urban fringes of the city (Redmond et al. 2012). Between 2000 and 2008, a sprawl pattern of development became established in the East Region of Ireland. A lack in provision of housing close to the economic core areas of the region created a continuing push of employment related

L. O. Petrov (✉)
The Executive Agency for Small and Medium-Sized Enterprises, European Commission, Brussels, Belgium

B. Williams · H. Shahumyan
School of Architecture, Planning and Environmental Policy, University College Dublin, Dublin, Ireland

© Springer Nature Switzerland AG 2019
D. Payne et al. (eds.), *Social Simulation for a Digital Society*, Springer Proceedings in Complexity, https://doi.org/10.1007/978-3-030-30298-6_12

housing demand at increasing distances from Dublin. This dispersal of housing, retail and employment activities in a fragmented manner across an ever-increasing area has major implications for the environment, infrastructure and service provision (Williams et al. 2010). Ireland has a flexible development led planning policy framework allied with direct local political power over the planning process and development control. This resulted in over-zoning and overbuilding of housing outside urban centres during the boom period. This trend combined with an underdeveloped public transit system within main urban centres such as Dublin resulted in extensive dispersed development leading the European Environment Agency in 2006 to describe Dublin as a worst-case example of sprawl. In 2009, the economy and development markets experienced a major downturn with all development activity stalled.

From 2007 to 2016, following the property market crash in Ireland, governments and policy processes were simultaneously addressing the banking and financial collapse which resulted from the property crash and attempting to restart the development process (Williams and Nedovic-Budic 2016). In the economic recovery from 2013 to 2018 the housing development industry has not yet recovered leading to emerging housing and property shortages in high-growth areas such as Dublin. Central Bank reports have noted that the financial implications of the property crash including non-performing loans fell with the recovery but the ratio remains among the highest in the euro area (Central Bank of Ireland 2015). It is evident that the problems of the environmental and economic consequences of the dispersed development activity during the boom will continue to have a major bearing on future development and environmental policies in the region.

Finding solutions for urban and regional policy problems is essential and a fruitful collaboration between scientists and stakeholders is crucial. This chapter relates recent research experience over the past decade in working with the stakeholders from national government, the Environmental Protection Agency (EPA), regional authorities and other bodies. This involved collaboration on the process of policy scenarios and selection of indicators in the GDR to support stakeholders on the future urban and regional planning and development decisions and strategies and subsequent related projects. This chapter provides an analysis of ongoing project practice, promoting cooperation between the various stakeholders and explores the relevant outcomes.

Methodological Approach

Scenarios and Indicators

The work carried out and presented here was developed from a seminar on Integration of Spatial Decision Support Systems and Evidence Based Modelling in National/Regional Policy Applications and Regulatory Systems: Scientific and Policy Challenges, at University College Dublin, on 9–12 July 2013, Dublin, Ireland and updated by discussions with stakeholders in 2016/2017. The selected stakeholders were engaged in the form of knowledge provision, scenario development, interpretation of results, and development of policy alternatives. They were actively involved in all stages of the work.

For the development of four realistic scenarios, the scientists collected information and proposed several narratives to policy makers (for transparency and clarity in the presentation the data/information and sources used was included also). In order that the stakeholders could understand based on what data and information the scenarios were built, they contributed to improve the scenarios, underlining the key drivers (main activities and main priorities in the GDR). This demonstrated which elements are most important to stakeholders and which information was not of immediate priority due to short-term needs and long-term strategies. The discussion between the groups facilitated the selection of the most relevant scenarios to be further translated into future land use maps. A detailed description of the scenarios "Business as usual" (BU), "Compact development" (CD), "Managed dispersed" (MD) and "Recession" (R) can be found in Petrov et al. (2011, 2013) and Shahumyan et al. (2011).

It is important to reflect on the current economic situation and the policy options for regional planning and development, which are the central driving elements of these scenarios. The BU scenario explores the further development of urban patterns emerging before the economic crisis whereas the R scenario focuses on future urban development due to recession, including a recovery by 2016, which did occur. The CD scenario is important in demonstrating how less pressure on natural land uses occurs with urban growth and regional development in the framework of a strong environmental protection policy. In the MD scenario was carried out a more detailed investigation of the permitting of growth and sprawl of hinterland towns and villages in open countryside, particularly along the Dublin-Belfast motorway. The realisation of this scenario is greatly facilitated by the planning regime, which imposes few constraints on the conversion of agricultural areas to low-density housing areas (Williams et al. 2010).

The use of modelling, in this case the MOLAND model is an added value for regional scenario policy discussions. The dynamic nature of the models had proved to be more helpful in the presentation of 'what if' rather than static models for policy and political interests. Therefore, the benefit of realizing scenarios over both spatial and temporal contexts was regarded as a major contribution. The use of a dedicated model based on changing economic activity and demographic trends levels activated algorithms which produce resulting land use patterns is clearly a useful addition to conventional land use transportation models which allocate travel volumes first and consequential land use impacts later. To conclude, choosing the right modelling is an important step for producing relevant results for stakeholders.

As a next step, a questionnaire was followed in 2016/2017 by structured discussions on important issues arising with the key stakeholders from the EPA, government departments and policy research officials from both the Dublin City and Regional Authorities. Four categories of indicators related to land use, urban growth/sprawl, and regional development for the regional planning and policy context of the GDR were discussed (Table 12.1). Further, only the indicators of 'high' interest are given in the results section.

Table 12.1 Relevance of indicators given by stakeholders

Indicators classification	Indicators	Group 1 stakeholders	Group 2 scientists
Land use indicators	Land use transformation vs time	High	High
Urban growth	Loss of natural areas vs. urban	High	High
Urban sprawl	Residential discontinuous vs. natural areas	High	High
Fragmentation	Total area	Low	Medium
	Number of patches	Medium	Medium
	Mean patch areas	High	High
	Total edge	Low	High
Connectivity	Euclidian nearest neighbourhood mean	Medium	Medium
	Mean min distance	Low	High

The stakeholders suggested also further indicators of high relevance for their decisions such as the degree of urbanisation, the population density, the employment density, the population pressure on agricultural/natural land, the population within 500 m of main road transportation axis. More indicators on quality of natural areas were also recommended.

Issues Discussed Between the Two Groups, Scientists and Policy Makers

Difficulties in the Collaboration Between Scientist's and Stakeholders

The stakeholders suggested that it was essential to involve them early in the process of building scenarios as well as at later stages in the work. The complexity of the land use theme was underlined, and the fact that the land use as a concept is not perceived as a real concern to stakeholders while however the resulting economic and residential activity are of major concern. Also, the relevance of reading the results in comparison with other 'local' land use typologies, historical, etc. was highlighted. In addition, it was stated that it would be helpful to know what comparable regions are doing how regions share information in an efficient way. The scientists principal concerns were to understand better where and how exactly their results are used by stakeholders as well as how their results are selected/validated during the policy process.

Improving Collaboration

It was suggested that the collaboration between the two groups can be improved by early involvement of stakeholders, building demonstration models and using them as a first step in the collaboration. In addition providing solutions for the real problems identified by policy-makers, rather than imposing methodologies in search of a problem was advised. This involves listening to policy-makers to understand their real problem and discussing the practical and pragmatic ways to assist policy-makers in using new tools and methods. More examples were needed as to how policy makers can read the scientific results both in individual cases and in comparison, with other related cases. It was regarded as essential that the scientific approaches should be closer to stakeholder's actual questions and priorities for example where they must take policy actions.

Obstacles for Applying Land Use Modelling Research for Practical Policy Making

Several relevant issues were mentioned: availability of data (and data at an appropriate scale); Lack of the required detailed land use data time series; Cost of building, calibrating and applying the required models; Lack of awareness of the potential end users; Lack of European legislation, framework, directive, criteria, and obligations on land use making individual regions unique and difficult to compare using standardised models.

Another obstacle is perception of the difficulty in using land use models in general. There is now much greater buy-in with regards to evidence-based planning but also understandable distrust of land use models by practitioners (planners) and politicians in instances where they do not understand the internal workings of the model.

Technical Aspects of Urban Models Used with the Stakeholders

The Cellular Automata-Based Model

The MOLAND model used in this study allows the exploration of urban dynamics generated by autonomous developments, external factors and policy measures by using 'what-if' scenarios. The model application was first developed by White and Engelen (1993), White et al. (1997) and later implemented by Petrov et al. (2009), Shahumyan et al. (2011), Ustaoglu et al. (2018) in several studies. It is applied for: urban development, regional development, transport corridors, adaptation to climate change driven hazards. More technical explanations of the model can be found in Barredo et al. (2003, 2004), Petrov et al. (2009), Shahumyan et al. (2014).

The model operates at three geographical levels: the global (one spatial entity, typically representing a country or group of countries, or an administrative or physical entity), the regional (administrative entities, typically represented at NUTS 2, NUTS 3 or NUTS 4 level) and the local (N cellular units). In this study, the global level represents the whole GDR. It is implemented based on the scenarios defined by the population and employment (industrial, commercial and services) projections entered in the model as global trend lines.

The regional level consists of five counties (Dublin, Meath, Kildare, Wicklow and Louth) as can be seen in Fig. 12.1. At this level the model requires socio-economic data for each of the modelled counties. For the simulation starting in 2006, the census population and employment data provided by the Irish Central Statistics Office (CSO) was used (more details in Sect. 12.2.3.2). For population in 2026, the CSO "M2F1 Traditional" projection was used for BU, CD and MD scenarios, assuming international migration declining with constant fertility and a return to the traditional pattern of internal migration by 2016; the CSO "M0F1 Traditional" projection was used for R scenario considering zero net migration and high fertility (CSO 2008). The model used allows classification of the total population of the region into two categories: "Sparse" and "Continuous". In the case of BU, MD and R scenarios, the same proportion of "Sparse" versus "Continuous" population in 2026 as was observed in 2006 was used. But, for CD scenario this

Fig. 12.1 The Greater Dublin Region highlighted in the map of Ireland

proportion was changed, making "Continuous" population in 2026 10% higher than in BU scenario.

Subsequently, at the local level, the global demands are allocated by means of a constrained CA-based land use model developed by White et al. (1997), White and Engelen (1993) evolving on 200 by 200 m grid. Land uses are classified in 23 classes, belonging to three categories: active (residential continuous dense, residential continuous medium dense, residential discontinuous, residential discontinuous sparse, industrial areas, commercial areas, public and private services), passive (arable land, pastures, heterogeneous agricultural areas, forests, semi-natural areas, wetlands, abandoned) and fixed (port areas, construction sites, roads and rail networks, airports, mineral extraction sites, dump sites, artificial non-agricultural vegetated areas, restricted access areas, water bodies) features (Engelen et al. 2007). The land use transition within a cell is determined by the combination of the following four local characteristics:

(a) The neighbourhood effect, which is the state of a cell at any time depending on the state of the cells within the neighbourhood. In our model, the neighbourhood effect represents the attraction (positive) and repulsion (negative) effects of the various land uses/covers within a radius of eight cells (the neighbourhood contains 196 cells). This iterative neighbourhood effect is found in the "philosophy" of standard CA, where the current state of the cells and the transition rules define the configuration of the cells in the next time step. An in depth explanation about this can be found in Petrov et al. (2009).

(b) A suitability map for each active land use. Particularly, for this study, to implement the effects described in each scenario, some modifications were made to the existing land use suitability maps for residential and industrial classes. The suitability of major or key towns was kept relatively high while in the rural areas it was decreased. Thus, the main differences of the suitability maps are: the highly suitable towns in Dublin-Belfast transport corridor for the MD scenario; and a restricted zone (2 km buffer along coastline) for future development for CD scenario. For the BU and R scenarios a default suitability map was used.

(c) A zoning map for each active land use, was also prepared in GIS. Here, the maps of the protected and conservation areas obtained from the Department of Environment, Local Government and Heritage (http://www.heritage.ie) were used to create a zoning map for all urban (activity) land use classes.

(d) The accessibility map. The accessibility for each active land use was calculated in the model relative to the road network consisting the motorways and main national and regional roads. The existing road and rail datasets were provided by the Dublin Transportation Office and the National Roads Authority. However, each one of the scenarios required different transportation links to be derived from the planned network. Transport 21 was a capital investment framework under the National Development Plan through which the transport system in Ireland was planned to develop, over the period 2006–2015 (www.transport21.

ie). It included seven light rails and two metro infrastructures by 2016.[1] Transport 21 was not included in MD scenario. Instead the accessibility of the current transport network was enhanced providing better connections to roads. In all other scenarios, Metro North was included as originally planned for introduction by 2014 and other Transport 21 railways by 2020.

Data Collection and Preparations

As mentioned above in the Cellular Automata-Based Model description, for the simulation at regional level, the census population and employment data provided by the CSO was used. Job numbers for BU scenario were extrapolated using population and job figures from CSO 2006 and CSO population projection of 2026, assuming that distribution of jobs in industry, commerce and services will remain in similar proportions as in 2006, and that the absolute numbers will increase in accordance to population growth by 2026. The MD scenario suggests a steady increase of economic growth by 2026. To implement this in the model, the employment projections of the BU scenario were increased by 10%. For CD scenario a projection of 30% more jobs in 2026 compared with MD scenario was used. The employment projections for R scenario are based on the figures of the Economic and Social Research Institute (www.esri.ie) where the recession is followed by a strong recovery (Bergin et al. 2009). The population and employment projections used in the simulations are presented in Table 12.2.

Model Calibration and Validation

The model calibration follows the approach proposed in White et al. (1997) and further developed in Barredo et al. (2003). The calibration method required running the model over a period for which actual and final land use maps are available.

Table 12.2 GDR population and employment projections used in the MOLAND model for scenario simulation

			2026			
		2006	BU	MD	CD	R
Population	"Continuous"	1,437,010	2,068,380	2,068,380	2,171,799	1,728,666
	"Sparse"	336,793	484,768	484,768	381,349	405,149
Employment	Industrial	259,800	380,573	418,630	544,219	300,603
	Commercial	321,790	559,519	615,471	800,112	372,329
	Services	211,656	362,129	398,342	517,844	244,898

[1] During the recession from 2019 to 2013 the Irish government decided to postpone many of the Transport 21 projects due to the economic situation in Ireland. However, in this study we run our scenarios with an assumption that with the economic recovery a number of the projects will be implemented by 2026.

These are typically a historical and the current land use maps. For the calibration of GDR, the land use maps of 2000 and 2006 were used. These maps were produced by ERA-Maptec Ltd as part of the Urban Environment Project (UEP 2006–2010, www.uep.ie). The simulated land use map of 2006 was compared with the actual land use map of 2006.

The calibration involves two further problems: (1) how to compare the output of the model with actual data and (2) how to define parameter values for improving the comparison results. These issues are also related to the validation processes (Engelen and White 2008) and are based upon two approaches: (1) visual comparison between the simulated land use map for the 2006 and the actual land use map for that year; (2) quantitative evaluation of the degree of coincidence between the two land use maps using the classical coincidence matrix and derived statistics as goodness-of-fit measures (Hagen 2003; Monserud and Leemans 1992).

Once the calibration and validation results was satisfactorily concluded, the future simulation of land use was carried out for a period of 20 years into the future. Further, the stochastic parameter, which has the function of stimulating the degree of stochasticity that is characteristic in most social and economic processes such as cities (Barredo et al. 2004) was set at 0.5.

The land use map of 2006 was the starting year of our scenario simulation (Petrov et al. 2011, Fig. 12.2). The calculation of land demand was associated with population and employment projections as well as the accessibility to transport networks (existing and planned roads). The suitability and zoning maps were also used according to the characteristics of the generated scenario. The level of stochasticity varied between scenarios because of the more scattered ($\alpha = 0.8$) or compact ($\alpha = 0.5$) distribution of land use purposes.

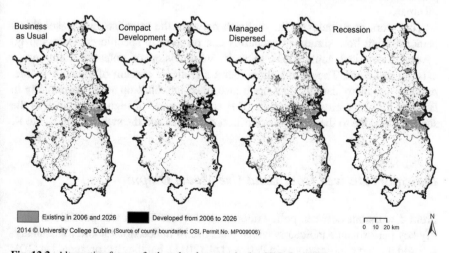

Fig. 12.2 Alternative forms of urban development in the GDR for 2026

Results and Discussion

Land Use Scenarios Modelling and Simulations

For policy makers it was clear that the results required covered both the outcome of the plan as projected and an evaluation of alternative scenarios which might arise or be considered. The scenarios used in this work incorporate policies or strategies which from discussions with policy interests had a realistic probability of being implemented. They also explore what could happen under circumstances outside of the direct control of stakeholders and decision makers. For example, in the MD scenario the transport corridor Dublin-Belfast is one of the key drivers determining a more scattered and diffuse urban land use pattern especially towards Northern Ireland while the simulated CD scenario presents a more compact urban land use pattern. This could be the result of a series of spatial planning regulatory measures and policies as well as due to the prevalence of the high-tech IT sectors and other economic activities which encourage a reduced distance travel to work. The R scenario was based on assumptions of the ESRI and is a result of stakeholder's feedback concerning the recession, economic recovery and ability to respond to better urban and regional planning and development (Fig. 12.2).

The research group provided an evaluation of the likely outcomes of alternative urban development approaches in terms of the form of urban expansion expected given different policy decisions. For example expansion and infrastructural changes were expected in the region due to the dominant role of the GDR in Ireland and the transport corridor Dublin-Belfast which was identified as a core axis on the East Coast. The results show an important issue for the GDR where new urban development, especially in Meath, Wicklow, Kildare and Louth, must comply with the best practices of sustainable development, protecting traditional agriculture and natural areas.

This work contributes to the dialogue between the scientists and stakeholders demonstrating how simulated scenarios can respond to specific contexts and go beyond theoretical applications to assist in provision of potential solutions to existing problems. The results show that a significant amount of valuable arable, pastures and many natural areas are transformed into built-up areas, particularly in BU and MD scenarios. The results of R scenario were of high importance to the stakeholders due to the pressures of economic cycles and the switch from BU to R.

Indicators Derived from Land Use Model Output

From discussions between policy interests and researchers it was possible to identify key performance indicators which were useful in decision making. This analysis used the same indicators as in Petrov et al. (2011), but they are presented at GDR level in Table 12.1 with comments on key issues arising following and were high-lighted by the stakeholders as of high relevance.

Land Use Transformation Versus Time Indicator

The land take was defined as the "change of the amount of agriculture, forest and other semi-natural and natural land taken by urban and other artificial land develop-ment" (European Environment Agency 2013). A key concern was and remains how fast the use of land is changed or even transformed irreversibly. The 'land use trans-formation versus time' is an indicator of land use change over a time period, involv-ing comparisons of each cover type for each cell between the start and end of the model run. The overall urban areas have grown considerably in all four active land uses (residential continuous, residential sparse, commercials and industrial). Rapid growth is foreseen for residential continuous in County Wicklow and for residential sparse in County Meath. Significant growth can be noticed in industrial areas in County Kildare, for commercial space in County Louth and services in County Meath. In summary during the 20 years simulated, the growth in urban areas popu-lation and economic developments which occur mainly in Meath, Kildare, Wicklow, and Louth counties.

As we know the natural areas inevitably shrink as urban areas grow (Kasanko et al. 2006). Therefore the significant growth of urban areas in the GDR led us to the intuitive indicator: indicating how much of the natural areas are taken by urban uses and development: where detailed analysis (Table 12.3) reveal that between 2006 and 2026, the loss of natural area (arable, pastures, heterogeneous agricultural, for-ests, semi-natural and wetlands) range from 3% in R scenario to 6% in MD sce-nario; details show that major loss occurred in County Dublin, followed by Louth, Kildare, Wicklow and Meath. Comparing the 2026-year scenarios, the heteroge-neous agricultural, arable and pastures are mostly transformed into urban areas in BU, CD and MD scenarios while pastures are the highest effected in R scenario. Nonetheless, it is notable that the loss of natural areas is the lowest in R scenario. For example, in County Dublin the loss of natural area is 10% in R scenario while it is 27% in CD scenario. Furthermore, the loss of heterogeneous agricultural and pasture areas is mostly in Dublin and Kildare counties (the maximum loss taking place in MD scenario of County Kildare) (Table 12.3).

Table 12.3 Urban areas and loss of natural areas from 2006 to 2026

Urban areas vs. natural areas	Scenarios (%)			
	BU	CD	MD	R
Total urban areas in GDR	86.2	89.2	89.3	51.8
Loss of total natural areas in GDR	−5.6	−5.8	−5.8	−3.4
Total urban areas in Dublin Co.	41.8	53.4	45.6	20.8
Loss of natural areas in Dublin Co.	−21.1	−26.9	−23	−10.5

Urban Growth Indicator

The 'urban growth' indicator as implied by its name is an inherently dynamic spatial phenomenon. Reasons for urban growth are manifold and can be identified with rising living standards (more space per person), developing commercial and transport services (which requires more buildings), changing living preferences (single houses preferred instead of blocks), changing land use policies, etc. This indicator aids in assessing the success or failure of growth management policies designed to control sprawl and/or protect specific natural lands (Kline 2000; Nelson 1999).

Comparing the 2026-year scenarios of GDR, the maximum urban growth occurs in County Meath in all scenarios, with a peak of 269% in MD scenario. Wicklow and Kildare counties also experience large urban growth. Residential continuous growth reaches a maximum in County Wicklow whereas residential sparse dominates the changes in County Kildare. In County Dublin, the urban growth is much lower, ranging from 121% in R scenario to 153% in CD scenario (Table 12.4). Nevertheless, the results of the CD scenario suggest that restricting the future development to within 2 km from the coastline will help to protect sensitive zones even in the face of increasing development.

Furthermore, the service activity areas present the highest difference of 290% (comparing to all active classes) between CD and R scenarios in County Louth. In the R scenario, the changes in the GDR are mostly relate to commercial and service activity areas; counties Meath and Louth are seen to be the most affected by the recession (Table 12.5).

Table 12.4 Urban growth for GDR from 2006 to 2026

Counties	Growth by scenarios (%)			
	BU	CD	MD	R
Louth	216.5	223.4	218.0	173.5
Meath	267.9	253.0	269.4	207.2
Dublin	141.8	153.4	145.6	120.8
Kildare	243.7	234.8	246.8	191.1
Wicklow	256.5	241.2	259.0	203.1

Table 12.5 The range between highest (BU/CD/MD) and lowest (R/CD) growth in the scenarios in percents

Active land uses	GDR	Dublin	Kildare	Wicklow	Meath	Louth
Residential continuous	36	67	24	102	262	194
Residential sparse	57	70	81	152	236	234
Industrials	90	24	41	64	75	79
Commercials	136	63	67	164	215	191
Services	110	81	58	125	218	290

Table 12.6 Urban sprawl in GDR

	Actual	By scenarios			
	2006	BU	CD	MD	R
Discontinuous residential area	100%	171%	114%	171%	127%
Ratio of discontinuous residential area and natural area	58	20	26	20	25

Urban Sprawl Indicator

Urban sprawl is expressed mainly through strong growth of sparse residential areas (EEA 2006). Kasanko et al. (2006) in their study showed that the city of Dublin had the highest rate of sparse residential areas among the 15 European cities studied. Our model output for the GDR projected land use indicates that compared to 2006 in 2026 the sparse residential areas may increase from at least 14% in case of CD, to 71% in case of BU and MD scenarios (Table 12.6).

Urban and Regional Development Indicators

The distribution of urban areas by counties is changing between 2006 and 2026. The County Dublin has the largest urban areas in all scenarios, but by 2026 its proportion decreases substantially compared with 2006 values. On the contrary, the proportions of Kildare, Meath and Wicklow counties increase over the period 2006–2026.

These results support the stakeholders in understanding the future trends of land uses of each county based on their decisions/strategies outlines for short and long term, different needs and local conditions whilst simultaneously being influenced by the key drivers of the whole region. They assist in targeting residential areas and activities for 'smarter' policy interventions within each county and assist in developing closer collaboration between counties for reaching harmonious outcomes for mutual benefit at county and regional levels.

Conclusions

This research clearly indicated that the stakeholder's focus was on indicators of high relevance for their current and priority decisions as they reflect better policy and practice priorities. Therefore, for the continuing collaboration between the research team, EPA and the regional Authorities, it is intended that more indicators will be produced together with updated scenarios to the extended periods of 2030/2050. This will support their strategic planning and development of the GDR, including providing the best possible long-term access to the modelling, including institutional host arrangements and the ability to disseminate key results.

The experiences of collaboration between scientists and stakeholders indicate that scientists tend to view research as a longer-term commitment whereas local policy decision makers have short term priorities and demands. This means that for stakeholders the process of engaging with research must focus on key priority issues and deliver results quickly while researchers focus is on the long-term repeatability and scientific rigour of the research. The urban development policy area presents both short and long-term policy decisions with a clear need for a scientific evidence base. The use of open sources of data and open modelling systems present significant opportunities for further research collaboration. The lengthy term commitments required in this process mean that short licensing periods for dedicated proprietary type models present significant limitations in creating viable research partnerships.

Acknowledgements The authors would like to acknowledge the support of the Environmental Protection Agency under contract 2005-CD-U1-M1. All work undertaken with the MOLAND model for the GDR is under licence from RIKS b.v. and Joint Research Centre under licence no. JRC.BWL.30715.

References

Barredo, J. I., Demicheli, L., Lavalle, C., Kasanko, M., & McCormick, N. (2004). Modelling future urban scenarios in developing countries: An application case study in Lagos, Nigeria. *Environment and Planning B: Planning and Design, 32*, 64–84.

Barredo, J. I., Kasanko, M., McCormick, N., & Lavalle, C. (2003). Modelling dynamic spatial processes: Simulation of future scenarios through cellular automata. *Landscape and Urban Planning, 64*(3), 145–160.

Bergin, A., Conefrey, T., Fitzgerald, J.D.,Kearney, I.M., (2009). Recovery Scenarios for Ireland. Research Series Number RS007, Economic and Social Research Institute: Dublin, Ireland.

Central Bank of Ireland. (2015). *Macro prudential review*. Dublin: Central Bank of Ireland.

Central Statistics Office (CSO). (2008). *Regional population projections 2011-2026, Dublin*. http://www.cso.ie.

Couch, C., Leontidou, L., & Petschel-Held, G. (2007). *Urban sprawl in Europe*. Oxford: Blackwell.

Engelen, G., Lavalle, C., Barredo, J. I., van der Meulen, M., & White, R. (2007). The MOLAND modelling framework for urban and regional land-use dynamics. In E. Koomen, J. Stillwell, A. Bakema, & H. J. Scholten (Eds.), *Modelling land-use change, progress and applications* (pp. 297–319). Dordrecht: Springer.

Engelen, G., & White, R. (2008). Validating and calibrating integrated cellular automata based models of land use change. In S. Albeverio, D. Andrey, P. Giordano, & A. Vancheri (Eds.), *The dynamics of complex urban systems* (pp. 185–211). Heidelberg/New York: Springer, Physica-Verlag.

European Environment Agency. (2006). *Urban sprawl in Europe, the ignored challenge, EEA report no 10*. Luxembourg: Office for Official Publications of the European Communities. ISSN 1725-9177. http://www.eea.europa.eu/publications/eea_report_2006_10/eea_report_10_2006.pdf.

European Environment Agency. (2013). *Land take*. Available on the Internet at http://www.eea.europa.eu/data-and-maps/indicators/land-take-2/.

Gkartzios, M., & Scott, M. (2005). *Countryside, here I come: Urban rural migration in the Dublin City-region* (Planning and Environmental Policy Research Series PEP 05/01). Dublin: University College Dublin.

Hagen, A. (2003). Fuzzy set approach to assessing similarity of categorical maps. *International Journal of Geographic Information Science, 17*(3), 235–249.

Hurley, J., Lamker, C. W., Taylor, E. J., Stead, D., Hellmich, M., Lange, L., Rowe, H., Beeck, S., Phibbs, P., & Forsyth, A. (2016). Exchange between researchers and practitioners in urban planning: Achievable objective or a bridge too far?/the use of academic research in planning practice: Who, what, where, when and how? *Planning Theory & Practice, 17*(3), 447–473.

Kasanko, M., Barredo, J. I., Lavalle, C., McCormick, N., Demicheli, L., Sagris, V., & Brezger, A. (2006). Are European cities becoming dispersed? A comparative analysis of 15 European urban area. *Landscape and Urban Planning, 77*, 111–130.

Kline, J. (2000). Comparing states with and without growth management: Analysis based on indicators with policy implication comment. *Land Use Policy, 17*, 349–355.

Monserud, R. A., & Leemans, R. (1992). Comparing global vegetation maps with the kappa statistic. *Ecological Modelling, 62*, 275–293.

Nelson, A. C. (1999). Comparing states with and without growth management: Analysis based on indicators with policy implications. *Land Use Policy, 16*, 121–127.

Petrov, L. O., Lavalle, C., & Kasanko, M. (2009). Urban land use scenarios for a tourist region in Europe: Applying the MOLAND model to Algarve, Portugal. *Landscape and Urban Planning, 92*, 10–23.

Petrov, L. O., Shahumyan, H., Williams, B., & Convery, S. (2011). Scenarios and indicators supporting urban regional planning. *Procedia – Social and Behavioral Sciences, 21*, 243–252, Elsevier.

Petrov, L. O., Shahumyan, H., Williams, B., & Convery, S. (2013). Applying spatial indicators to support sustainable urban futures. *Environmental Practice, 15*(1), 19–32.

Redmond, D., Williams, B., Hughes, B., Cudden, J.: Demographic trends in Dublin. Dublin City Council: Think Dublin! Research Series: Office of international relations and research (2012).

Shahumyan, H., White, R., Petrov, L. O., Williams, B., Convery, S., & Brennan, M. (2011). Urban development scenarios and probability mapping for greater Dublin region: The MOLAND model application. In *Lecture notes in computer science (LNCS)* (Vol. 6782, Part 1, pp. 119–134). Berlin, Heidelberg: Springer.

Shahumyan, H., Williams, B., Petrov, L., & Foley, W. (2014). Regional development scenario evaluation through land use modelling and opportunity mapping. *Land, 3*(3), 1180–1213.

Ustaoglu, E., Williams, B., Petrov, L. O., Shahumyan, H., & van Delden, H. (2018). Developing and assessing alternative land-use scenarios from the MOLAND model: A scenario-based impact analysis approach for the evaluation of rapid rail provisions and urban development in the greater Dublin region. *Sustainability, 10*(1), 61. https://doi.org/10.3390/su10010061.

White, R., & Engelen, G. (1993). Cellular automata and fractal urban form: A cellular modeling approach to the evolution of urban land-use patterns. *Environment and Planning A, 25*, 1175–1199.

White, R., Engelen, G., & Uljee, I. (1997). The use of constrained cellular automata for high-resolution modelling of urban land-use dynamics. *Environment and Planning B: Planning and Design, 24*(3), 323–343.

Williams, B., Hughes, B., & Redmond, D. (2010). *Managing an unstable housing market* (UCD Urban Institute Ireland Working Paper Series 10/02). Dublin: Urban Institute Ireland, University College Dublin.

Williams, B., & Nedovic-Budic, Z. (2016). The real estate bubble in Ireland: Policy context and responses. *Urban Research & Practice, 9*(2), 204–218.

Chapter 13
Synthesising the Geography of Opportunity in Rural Irish Primary Schools

Gillian Golden

Introduction

There is an increasing recognition of the need to "complexify" the approach to many societal challenges, which often seem to remain intractable regardless of resources and reform efforts invested. One such issue is social inequality, which begins as an accident of birth and compounds over the lifecycle to produce increasingly polarised outcomes for different groups within the population (Heckman 2008). Since the global economic crisis began in the mid 2000s, it has been well-recognised that despite economic progress, current public policy approaches for creating more equitable societies appear to have delivered limited success in many countries and a more sophisticated, individualistic approach is indicated to understand and ameliorate this state of affairs (OECD 2016).

The set of characteristics of systems and processes which are grouped in the academic literature under the umbrella terms of "complex systems" or "complexity theory" are concerned with understanding how individual circumstances and behaviours of units within systems feed into organising and generating subsystem and system level states. For the policymaker, analysing social systems through a complexity theory lens offers a departure from the "world as machine" view of policy intervention, where the central belief is with enough analysis and careful choice of inputs, the outcomes on the system can be controlled, towards a "world as organism" view that social systems adapt and change as the environment around them evolves (Innes and Booher 1999).

Complexity theory might then intuitively appeal to those that seek to have greater insight into the systems which they are charged with improving. However for the policymaker, moving from recognising this appeal to practical application in a real-

G. Golden (✉)
Dynamics Lab, University College Dublin, Dublin, Ireland
e-mail: gillian.golden@ucdconnect.ie

© Springer Nature Switzerland AG 2019
D. Payne et al. (eds.), *Social Simulation for a Digital Society*, Springer
Proceedings in Complexity, https://doi.org/10.1007/978-3-030-30298-6_13

world setting is a difficult transition. Social simulation offers a bridge. Goldstone (2006) argues that simulation promotes deep understanding of a system compared to mathematical representations as it meshes well with people's own mental models of systems; the structure of model elements and the forces that cause interactions between them and promote specific outcomes are "close to natural psychological structures and processes". The field of social simulation has developed and matured significantly in the last 15 years and has established itself as a distinct research discipline (Meyer et al. 2009; Hauke et al. 2017).

In tandem, individual level models of social systems are becoming more popular, as computational power and methodological capacities expand. Many classes of such models exist, such as microsimulation models, which simulate the impact of policy change on individual micro units within systems (Harding 2007) and agent-based models, which are composed of individual agents which have autonomous decision making abilities and can interact with each other and adapt to their environments (Macal and North 2010). However, regardless of whether a researcher intends to operationalise an agent-based model, a microsimulation model, or perhaps a hybrid, the process begins with the creation of a base dataset, which can often involve specifying a synthetic population, defined neatly by Harland et al. (2012) as *"a population built from anonymous sample data at the individual level, which realistically matches the observed population in a geographical zone for a given set of criteria"*.

Where synthetic populations are created with a spatial component, a common construction technique is to generate a "best set" of synthetic individual records according to census geographic block totals for the variables of greatest importance or interest for the population and micro-data samples from census data or a relevant survey which act as seeds to construct the population (Ballas et al. 2005b). Methods used to select records could be deterministic (such as, an algorithm which weights and selects records in the micro-data sample based on an estimated joint distribution of the specific attributes of relevance for the population generation) or stochastic (such as, begin with a random selection from a candidate set of records and iterate by randomly exchanging records until convergence within acceptable limits of the marginal distribution of the variables of interest is reached).

However, the methodology in this field is rapidly evolving. There is a growing imperative on national statistical organisations and other gatekeepers of public data to make more anonymised data available for research (see, for example, the manifesto of Helbing and Balietti (2011)) and to harness the statistical potential of administrative records. We are quickly moving towards an era of *"big microdata"* (Ruggles 2014) which will provide new avenues for researchers to access large quantities of microdata, either through partially synthetic full coverage files with in-built disclosure controls created by public authorities (Reiter and Mitra 2009; Abowd and Lane 2004) or where researchers use administrative data as a basis to generate realistic, tailored purpose synthetic populations (for example, Rich and Mulalic (2012)). In this spirit, the following sections describe the development and validation of a tailored synthetic population constructed from full-coverage anonymised census microdata, in order to examine the issue of educational risk prevalence in rural Irish primary schools. The intention is for this population to serve as the initial conditions for an agent-based model of educational risk that is broadly reflective of the actual situation in rural Irish schools, and which can be used to gain insight into the impacts of school and community-based interactions of individuals with different risk profiles, as well as test possible effects of different policy options to improve equity of opportunity.

The next section discusses the status quo of rural educational risk in Ireland and motivates the construction of the synthetic population. Section "Data" provides an overview of the data. Section "Description of the Population Generation Algorithm" describes the algorithm used to contruct the population, and section "Specifying Pseudo-Catchments" explains the process for creating school catchments for the population. Section "Incorporating Constraints" describes the use of a multi-objective combinatorial optimisation algorithm to select school populations from the pseudo-catchments. Finally the "Results and Discussion" section presents the results of the algorithm runs, along with a discussion of some validation issues and the features of the synthetically generated population.

Rural Educational Risk in the Irish Context

Disparities in educational attainment play a key role in determining the picture of inequality in a society as a whole. Spatial location is recognised as being one source of inequalities in educational outcomes (Tate and William 2008; Green 2015) and other types of inequalities, a phenomenon which has often been referred to as the "*geography of opportunity*" (Galster and Killen 1995). While debates on exactly what educational inequality means and the underlying mechanisms by which it develops and propagates are ongoing (see Boldt et al. (1998) for a detailed treatment in the Irish context), literature has identified certain factors which, although they do not guarantee poorer educational outcomes for young people, certainly greatly increase the risk of what Ainscow (1999) refers to as "*marginalisation, exclusion or underachievement*" in education. Therefore researchers and education policymakers are interested in identifying the prevalence and dispersion of these risk factors and designing targeted interventions to mitigate against them.

In Ireland, the government seeks to attenuate these risks by allocating extra resources to target educational disadvantage, which is defined in Irish law as "…*the impediments to education arising from social or economic disadvantage which prevent students from deriving appropriate benefit from education in schools.*" (Education Act 1998, IRL). The funding formula for Irish primary schools includes a need-based component which allocates some additional resources to schools based on a judgement of which schools have the highest proportions of these social or economic impediments in their student population. The most recent of these schemes is known as the Delivering Equality of Opportunity in Schools programme (DEIS) though similar programmes have been in existence in Ireland since the 1980s. Until recently, schools have been identified for inclusion in the DEIS scheme based on a survey completed by school principals which required them to estimate the proportions of pupils with certain risk factors (such as whether they are living in local authority housing, or whether they are a child of a lone parent family) in their schools. There is a tacit consensus that the most disadvantaged schools in urban areas are well identified, as the areas of large urban centres which are most impoverished are highly visible. However the process of identifying and understanding how educational risk presents and propagates in rural schools is not as clear-cut.

Urban disadvantage is prioritised over rural disadvantage in the DEIS scheme, with the majority of resources devoted to urban schools, though it has been estimated by Kellaghan (1995) that in fact 60% of all disadvantaged primary school pupils are living in rural areas (defined as areas with less that 10,000 population). Kelleghan openly acknowledged that rural areas *"receive inadequate attention under existing schemes"* and require more focused research. Since 1995 however, there has been no notable changes to the distribution process for DEIS resources between urban and rural schools. In some respects, the lack of urgency is understandable given available evidence; as a group, students in rural schools tend to either perform as well as, or better than their urban counterparts in various standardised tests. Weir et al. (2009) also find a distinct absence of a social context effect in rural Irish primary schools. However it has been demonstrated by Haase et al. (1995) that these aggregated statistics hide pockets of significant deprivation in rural areas, and it has also been argued as far back as the 1990s that relevant socio-economic variables from census data should be integrated into the identification process for educational disadvantage in Ireland (Haase 1994). The Irish Department of Education and Skills is in the process of modifying the resource allocation model for DEIS to take into account local socio-economic profiles as derived from census data (Department of Education and Skills 2017) however much investigation still needs to be carried out to ascertain how to meet the needs of those at risk of poor education outcomes in rural areas. This research aims to contribute to that investigative process.

Understanding rural disadvantage in Ireland is further complicated by the lack of a legally defined catchment-area system tied to official geographic boundaries (though many schools define parish boundaries as de-facto catchment areas). In addition, parents of primary school aged children have free choice of which school to send their child to, subject to the enrolment criteria set out by many schools to enrol students in the case of oversubscription. These facts hinder resource allocation schemes aimed at reducing educational risk which rely on measures of the area where the school is located, as pupils attending rural schools are likely to be more geographically dispersed than the immediate local area in many cases, and therefore *de facto* catchments are difficult to define. Though some research has been advanced into defining catchments and tracking movements of students within catchments (Pearce 2000; Parsons et al. 2000), there is not yet a substantive body of literature dealing with this question. Individualised data with residence and place of school information available from the census can shed light on de-facto catchment areas and their social profiles. In the absence of an actual dataset with all relevant variables, this project aims to construct one synthetically.

Data

For the purposes of this research, a number of publicly-held datasets were combined to generate a realistic synthetic population. Firstly, publicly available GIS boundary files for two different levels of census geography (Electoral Divisions and Small Areas) used in the Irish Census 2011 were downloaded from the Irish Central Statistics Office website. Though these boundary files are slightly "smoothed" compared to the official Ordnance Survey Ireland boundaries they are widely used and considered sufficient for this analysis. Electoral Divisions (EDs) are the smallest legally defined administrative areas in the state, number 3340 in total, while the 18,445 Small Areas (SAs) generally cover an area of 18–120 households, depending on population density. Secondly, a school level file was created by merging various school-level data available from the Irish Department of Education and Skills, including DEIS status of the schools, enrolment totals as at 30th September 2011, gender type of the school, DEIS status, and XY coordinates detailing the spatial location of the schools.

As discussed in the previous section, most synthetic populations tend to use individualised census data as seed data, generally at a level of 10% sample of the entire population. However, this work benefited from having a more comprehensive dataset of records from Census 2011, known as POWSCAR (Place Of Work, School and College Anonymised Records). POWSCAR is a full-coverage micro-dataset containing information on the social and economic circumstances, place of school or work and commuting patterns of all enumerated workers and students in Ireland during Census 2011, along with generalised spatial information. This dataset is a prime example of recent innovations by national statistical organisations in making high-coverage, partially synthetic and/or generalised data files available for research purposes, as discussed in the introduction of this paper. POWSCAR is made available only under strict access and security conditions to bona-fide researchers who are undertaking a relevant project in the social sciences. To ensure the anonymity of the data, many of the fields have been generalised or obfuscated. Further details and a detailed user guide for POWSCAR can be found at CSO (2013).

POWSCAR contains a significant number of variables which have been found in the literature to be associated with risk of educational disadvantage, including parental education levels, employment status of parents, family situations (including lone parent status), and living conditions. Annex 13.1 contains the full set of variables available from POWSCAR. Furthermore, the inclusion of spatial information on area of residence, along with "fuzzy" information on place of school allows for a comprehensive analysis of the relationship between educational risk variables and the geographic location of students and their schools. Therefore, the POWSCAR dataset offers novel and very significant possibilities for developing a realistic synthetic school population for spatial analysis of educational risk.

The POWSCAR file, though theoretically offering full coverage of the school population, is not without limitations, which means that it serves better as a seed file for generating a realistic synthetic population rather than being directly used for

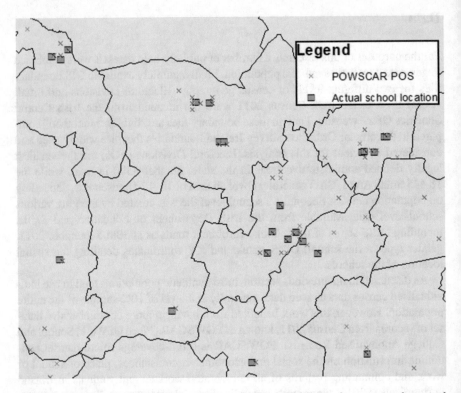

Fig. 13.1 Example of POWSCAR Fuzzy Place of School (FPOS) locations compared to actual school locations

school-level analysis of education risk. Firstly, though POWSCAR contains a rough geographic marker for the place of school for each record through the inclusion of a "fuzzy northing" and "fuzzy easting" of the place of school (which, for simplicity is referred to as "Fuzzy Place of School" and/or FPOS throughout this paper), the location was generalised enough not to allow for direct matching with specific school locations in all cases with sufficient levels of confidence, particularly in areas where there are two or more schools located very close together (see Fig. 13.1 for a mapped visual example). Furthermore, approximately 8% of the record set had a blank or uncodeable place of school (Table 13.1). Finally, the Department of Education National Annual Schools Census recorded a total of 509,652 (Table 13.2), compared to a total of 500,440 records of primary school students in POWSCAR. A number of factors explain this difference:

- The number of pupils recorded in the Department of Education and Skills Annual Census was based on head counts as at 30th September 2010, while the Census of Population took place in April 2011. Therefore there would have been some fluctuations in numbers during this time.
- The Census of Population was based on the enumerated population as at Census night i.e. those that were actually present in the country. Therefore any children

Table 13.1 Primary school records classified by high-level place of school, Census 2011

Place of school (POWSC location)	Frequency	Percent
Outside the state	61	0.0
Blank/uncoded	39,556	7.9
Home schooled	176	0.0
Ireland	459,750	91.9
Northern Ireland	897	0.2
Total	500,440	100.0

Table 13.2 Primary school official enrolment totals as at September 2010, as reported to the Department of Education and Skills

Gender	Enrolment totals
Girls	248,008
Boys	261,644
Total	509,652

of primary school age who are usually resident but not present in the State on Census night were not included.

- The designation of "primary" used in the POWSCAR data is based on the age of the student rather than the school that they are attending, so therefore all children aged 5–12 were assigned to be "Primary school children". In reality small numbers at either end of this age group would have been in primary school in April 2011, and small numbers of 12 year olds would also have already moved to secondary education.

As can be seen in Table 13.1, a number of records were coded to places of school outside of the State, or were reported as being homeschooled. As this study aims to generate a synthetic population for those attending primary schools in Ireland only, these records were removed from the analysis. While it is outside the scope of this report to motivate the decision process for risk variables, choices were made based on both availability within POWSCAR, and on existing literature on factors which have been strongly associated with greater educational risk. Derived flag variables were computed for three indicators for increased risk (lone parent family, living in rented accommodation, and where all parents were unemployed in the household) to be used as constraints in the population generation process (Table 13.3). Parental education is also strongly linked to educational risk in the literature. In the absence of sufficiently disaggregated information on low parental education, a "protective" factor was instead also added as a constraint, a flag for whether at least one parent had tertiary education.

Finally, as this analysis was concerned with rural educational risk, primary school students with places of school located in cities were removed. However, other large urban areas outside cities, though technically not rural (defined as per other Irish analyses such as Kellaghan (1995) as areas containing a population of less than 10,000), are retained in the population selection process in order to examine the patterns of risk both in large towns and their surrounding rural areas, which may be important for later identifying locations which would benefit most from intervention.

Table 13.3 Distributions of risk variables in urban and rural primary school populations, Census 2011

	Lone parent family				Rented accommodation				All parents unemployed			
	No		Yes		No		Yes		No		Yes	
	Count	%	Count	%	Count	Row N %	Count	Row N %	Count	Row N %	Count	Row N %
Rural (pop. <10,000)	237,243	86.3	37,763	13.7	221,948	80.7	53,058	19.3	31,875	84.4	5888	15.6
Urban (pop. >10,000)	169,915	75.8	54,385	24.2	136,465	60.8	87,835	39.2	44,483	81.8	9902	18.2
Total urban and rural	407,158	81.5	92,148	18.5	358,413	71.8	140,893	28.2	76,358	82.9	15,790	17.1

Description of the Population Generation Algorithm

Most spatially focused synthetic populations are generated using census data, where the summary characteristics of the geographic block are known and individual data is generated from some non-synthetic micro-data sample which statistically matches the proportions of characteristics of interest to the analysis when aggregated. Validation involves comparing the aggregated totals of the synthesised individual records with the non-synthetic actual totals for each block, and re-iterating the algorithm until the distances between observed and generated distributions of characteristics are minimised. However, this approach requires that the corresponding block totals are known for each of the areas in which we wish to generate a population from seed data (although in the absence of seed data, populations can also be generated solely using the aggregated data for each census block, as discussed in Harland et al. (2012) and Huynh et al. (2016)).

The data described in the previous section presents a different scenario. Here, the school enrolment totals and the spatial locations of schools are known, and the desire is to synthetically decompose these school enrolment totals into individual records containing information on risk factors. However unlike the situation with decomposition of census block, totals for these risk factors are <u>not</u> available in aggregated form in the school level and must instead by aggregated from the individualised data in POWSCAR. While within traditional synthetic models methodologies exist for assigning students to schools (Wheaton et al. 2009), these are generally underpinned by assumptions that the students attend schools in their geographic area subject to school capacity limits, with random selection thereafter to fill up any spare capacity in schools. This is a suitable approach for when the schools are simply acting as conduits for modelling large-scale collective gatherings (for example to simulate sites for large scale disease transfer in pandemic modelling, or significant concentrations of traffic at certain times of day), or in situations where school choice is known to be constrained to specific local geographic catchment areas. But when instead the specific socio-economic features of school and students are the key foci, and parents theoretically have free school choice, using a file such as POWSCAR provides a more realistic alternative as it is based on actual school enrolment patterns and has high coverage. A novel approach is therefore proposed for generating the school populations, which has two main steps and is conducted at both the school and Electoral Division level. The steps are:

- Specify a "pseudo-catchment" (in the form of a subset of POWSCAR records) for each school from which a set of records can be selected to represent the individual school enrolments.
- Use a multi-objective combinatorial optimisation algorithm (MOCO) to select records from the pseudo-catchment for each school in an electoral division which best maximises the "correctness" for the school population, and also maximise the "correctness" at the level of the electoral division of the specified flags for educational risk (lone parent family, rented accommodation, all parents unemployed) as well as the flag for tertiary education attainment of parents.

Specifying Pseudo-Catchments

A number of spatial pre-processing operations were carried out in order to prepare the datasets and to generate neighbourhood pseudo-catchment areas for each school. Firstly, as discussed in the previous section, records with a place of school that fell within city boundaries were "clipped out" of the dataset by overlaying the boundaries of the cities with the school locations. Secondly, for the different spatial operations, it was necessary to be able to plot all the POWSCAR records in GIS according to their place of residence as well as the fuzzy place of school location. As the POWSCAR file only contained information on the small area (SA) of residence of the students a random point was generated within the small area of residence for each student record so that spatial operations could be carried out.

To specify the pseudo-catchments, a number of approaches were tested. Initially a distance-based buffer was specified around each school that covered up to the 95% percentile of average distances from the random point of residence to the FPOS. However this led to very large catchment areas in more populated areas, and had implications for the later performance of the assignment algorithm. Further exploration showed that a set of records for each school which contained all those that had a FPOS within the same SA (but not necessarily living there), as well as all those that lived in the ED where the school was located gave more efficient and accurate performance later, as 92% of records had a specified SA for the place of school, and almost half of all students lived in the same electoral division as they attended school. These facts were therefore used to specify the initial pseudo-catchments. Safeguards were also included for gender to ensure that boys did not get assigned to girls-only schools and vice versa. Figure 13.2 outlines the decision process visually.

Incorporating Constraints

As the distribution totals for the socio-economic variables are not available at the school level, specifying the constraints required additional consideration. Marginal tables could however be calculated for the places of school at the small area level (SA) and at the electoral division (ED) level by directly aggregating the data available in POWSCAR by the Place of School locations, bearing in mind that, as discussed earlier, these tables were subject to a proportion of records (about 8% on average) having missing data as Place of School was blank or uncodeable for these values. The data aggregated to the ED level was used as the basis for constructing the constraining tables.

Given the use of an already computationally intensive MOCO to assign student records to schools (as described below), it was important that the constraints were chosen to maximise realism while also allowing for a feasible runtime for the algorithm. Following some experimentation with running the algorithm with various constraints, the following constraints were defined.

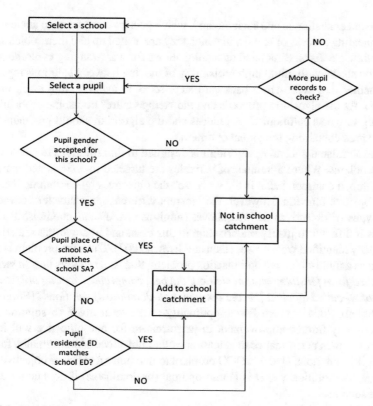

Fig. 13.2 Decision flow diagram for creating pseudo-catchments for each school

- The selected set should have the maximum possible proportion of records with the fuzzy "Place of School" marker in the immediate local area of the school location (i.e. in the same SA). This statistics is denoted throughout the remainder of this paper for brevity as P-FPOS (Proportion of records with matching Fuzzy Place of School at the Small Area level). Individual records were flagged with a value of 1 if they met this criterion and 0 otherwise.
- The mean absolute error (MAE) for the key risk indicators within each ED should be minimised. The MAE is defined as:

$$MAE = \frac{1}{n}\sum_{i=1}^{n}\left|y_i - \hat{y}_i\right|$$

where y_i is the actual proportion of records in each electoral division for risk indicator i (according to the marginal tables constructed for each ED), and $\hat{y}_i$ is the proportion according to the selected set of synthetic records for the ED. The MAE was calculated across the three risk indicators, as well as the indicator on tertiary attainment of parents.

Combinatorial optimisation methods have become commonplace for generating synthetic populations (Ballas et al. 2005a; Hynes et al. 2009; Nakaya et al. 2007).

Simulated annealing is one such method, which gets its name from an analogy with the annealing process of metals, in which they are heated up and then cooled slowly to minimise defects. Simulated annealing allows for wide-ranging exploration of a solution space, with initial high probability of moving between solutions regardless of whether they represent a backward step, so as to avoid getting stuck in local optima. As the algorithm proceeds and the "temperature" cools, the probability of moving to worse solution states reduces and the algorithm should gradually focus on the area containing the global optimum.

The simulated annealing algorithm is designed to find solutions to single objective problems. Where a number of variables are targeted in a simulated annealing algorithm, a common tactic is to "scalarise" the objectives by combining them into one objective function. However, the inherent difference in character between the two types of objectives defined above, (minimise MAE and maximise P-FPOS) makes it difficult to justify scalarisation in this case, and in any event scalarisation has many identified weaknesses (Duh and Brown 2007). Thus, the problem is maintained as a multi-objective optimisation problem, that is, it requires that "*a search is launched for a feasible solution yielding the best compromise among objectives on a set of so-called efficient (Pareto optimal, non-dominated) solutions*" (Ehrgott and Gandibleux 2004), where Pareto-optimality is the term given to solution states whereby any further improvement in performance for one objective will lead to deterioration in performance in at least one other objective. More formally, for a set of feasible solutions D to a MOCO problem to maximise a series of objective functions $f_i(x)$, a solution $x \in D$ is Pareto-optimal (or dominates) if no solution $x' \in D$ exists such that

$$\forall j, f_j\left(x'\right) \geq f_j\left(x\right) \text{ and for at least one } j \; f_j\left(x'\right) > f_j\left(x\right)$$

The set of Pareto optimal, non-dominated solutions is called the efficient frontier or the Pareto front. For the generation of the synthetic school population, this paper adopts a procedure outlined by Czyzak and Jaskiewicz (1998), which extends the simulated annealing algorithm to deal with multiple objectives. In this algorithm separate objective functions are maintained and an initial set of random solutions is created over the solution space, known as the generating set. During the run the algorithm perturbs this generating set to find new solutions and in some cases moves to the new solution and searches from there in the next iteration. In this way it finds a set M of non-dominated solutions (i.e., no member of the set is dominated by any other member) which approaches the Pareto front. The set M is updated with a new solution X by checking X against existing solutions in M. If X is non-dominated by these solutions it is added to M. At the same time if it dominates other solutions in M they are removed. A weighting vector Λ^x is used on the objectives and weights are rebalanced at each iteration and included in the calculation of the probability of moving to each new solution. The weights are calculated with respect to the nearest non-dominated neighbour solution, in order to "push it away" and ensure that the generating solutions are spread across the entire solution space. The probability of moving from one solution to another is linked to the temperature, the values of the weighting vector and the differences between the values of the objective functions. Specifically, (letting α be a constant close to 1):

$$P\left(x,y,T,\Lambda^{x'}\right) = \min\left\{1,\exp\sum_{j=1}^{D}\lambda_j^x\left(f_j\left(x\right)-f_j\left(y\right)\right)/T\right\}$$

where, for the closest solution X' to X

$$\lambda_j^{x'} = \begin{cases} \alpha\lambda_j^x \\ \lambda_j^x / \alpha \end{cases}$$

The weighting process is therefore set so that weights on objectives for which X is better than X' are increased and weights for which X is worse than X' are decreased. Figure 13.3 shows the algorithm applied to the pseudocatchments. As the algorithm proceeds, the probability of moving from the current solution drops. The algorithm outputs a set M of non-dominated solutions from each ED. As the objectives are considered equally important in the solution, a solution is then chosen at random from the set M from each ED to represent the final solution.

Results and Discussion

Simulated annealing is computationally expensive, and multi-objective simulated annealing even more so. The algorithm run-time increased significantly according to the number of generating solutions chosen (denoted r), and, to a lesser extent, slowing the cooling schedule. The starting temperature T was set to 10, with different cooling schedules tested. The weights were initialised at 0.5 for each objective, and α was set to 1.05. Different values for r were chosen, from 5 to 50 on some test EDs, however for anything greater than $r = 25$, the algorithm essentially became unrunnable for the entire set of EDs. In general, the algorithm showed marginal improvements in performance for greater values of r, while slowing the cooling schedule resulting in similar improvements for a lower computational cost. Some examples of the objective function values for computed sets of M are given below for ED 01001 for initial test runs of the algorithm, using different values for r and different cooling schedules as seen in Figs. 13.4, 13.5, and 13.6.

During initial runs a large variety in performance was observed between EDs, which was traced to the relative size of the pseudo-catchments in some areas compared to the numbers of records being sampled (i.e. where school enrolment numbers were small). A solution was found in the form of a "shrinking" operation to make the size of the pseudo-catchments more dynamic to the numbers of records sought, while maintaining all records where the place of school was known to be in that area. This amounted to, for each pseudo-catchment, selecting all records with a FPOS in the Electoral Division and a further random amount of records equal to 1.5

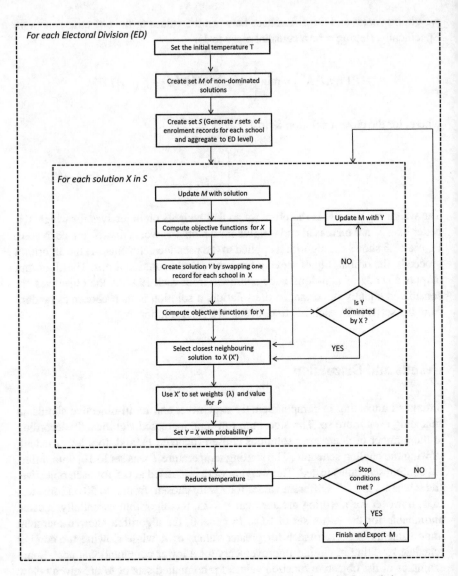

Fig. 13.3 Multi-objective simulated annealing process for generating school populations

times the total enrolments in schools located in the electoral division, and removing the remainder of records from the catchment. This led to a significant immediate improvement in performance as seen in Fig. 13.7.

Validation of synthetic populations created using combinatorial optimisation techniques can be problematic, and there is no generally accepted method of assessing goodness of fit, especially as goodness-of-fit cannot be considered to be absolute, but relative to the purpose of the population being simulated (Voas and Williamson 2001). As the random samples initially drawn from the pseudo-catchment areas are then optimised according to the specific purposes for generating

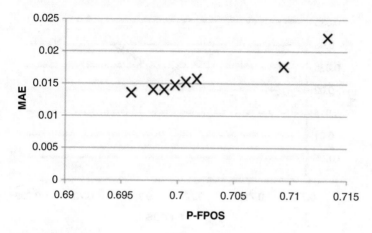

Fig. 13.4 MAE and P-FPOS (r = 5, T = T − 0.1)

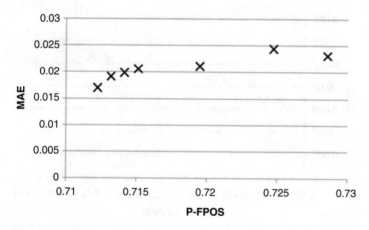

Fig. 13.5 MAE and P-FPOS (r = 15, T = T − 0.1)

the population, conventional statistical analysis tools cannot be applied to the resulting estimates, as pointed out by Voas and Williamson (2000). In the same paper Voas and Williamson also state that if possible when fitting data to different geographic blocks, to set an acceptance threshold for each block to ensure that uniformly satisfactory results are obtained. However they also acknowledge that this will lead to an unpredictable computation time. For the computational load involved in this algorithm, requiring a minimum quality threshold (for example P-FPOS > 0.7 and MAE < 0.1) could only have come by "exiting" an ED from the process once this threshold was hit to allow the computational resources to transfer towards more intensive searching of solution spaces for EDs that had not yet met the threshold. It was then unclear how much the benefit of having all EDs reach the minimum threshold would have been offset by the lost opportunity of having many of the EDs

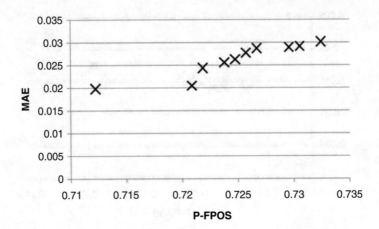

Fig. 13.6 MAE and P-FPOS (r = 10, T = T − 0.05)

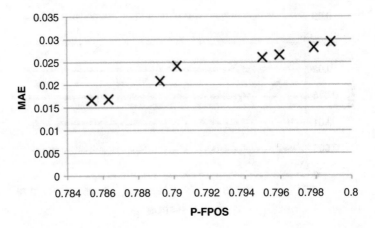

Fig. 13.7 MAE and P-FPOS (r = 10, T = T − 0.05, shrunk catchment)

achieve a higher quality solution than the minimum threshold. The following figures give visual views of the values for all EDs for each of the objective functions (MAE and P-FPOS) and a scatter plot of the values for each ED in the final synthetic population.

The general approach for computing the goodness of fit for synthetic populations is to compare the absolute or proportional values for the marginal tables in the synthetic data against the actual data. The Freeman-Tukey statistic can be used to calculate some measures of goodness of fit. It is defined as follows:

$$FT^2 = 4\sum_i \sum_j \left(\sqrt{S_{ij}} - \sqrt{A_{ij}} \right)^2$$

where S_{ij} is the ijth cell from the synthetic table and A_{ij} is the ijth cell from the actual table. The synthetic population is considered to be a good fit if the FT^2 does not

Table 13.4 Fit on constraints and partially constrained tables

Variable	Average FT^2	% EDs exceeding critical value (N = 1842)
Constrained flag variables		
Lone parent status	0.747	0.9
Rented accommodation	0.946	1.6
Partially constrained/unconstrained variables		
Sex	0.989	0.9
Nature of occupancy	2.503	3.5
Education level of parents	2.556	5.1
Household composition * family employment status	5.086	1.5
Nationality * sex	4.456	2.2

exceed the 5% χ^2 critical value with the degrees of freedom equal to the number of cells in the table. Marginal tables were aggregated for each ED separately for some constrained, partially constrained and unconstrained variables (according to the place of school for each record) and compared with the same tables aggregated for the synthetic population. Proportional differences were used for the calculation, in order to minimise the impact of the 8% with missing values for Place of School, although it is also acknowledged that due to the large variety in the sizes of the school population generated across EDs, proportion errors were magnified in smaller EDs even though the actual numerical difference in numbers of students with particular risk factors may have been small.

As can be seen from Table 13.4, the synthetic populations for a small proportion of EDs did not fit the constraint flag variables well, and these were also readily identifiable from a scatter plot of the objective function values for all EDs in the simulated population (Fig. 13.8). Further examination of the data revealed that the majority of the EDs with poor fit had very large FT^2 values which were attributable to geocoding issues, for example where a large proportion of records were geocoded to a place of school in the wrong ED, or at the edges of cities where the outer city boundary overlapped with a portion of an Electoral Division, causing inconsistencies in the results for the actual and synthetic population (see Fig. 13.9 for an example at the edge of Limerick City). However, overall the synthetic population can be considered to be a good fit, though manual investigation and resolution of the small amount of geocoding problems, future experimentation with further adjustments to the pseudo-catchment areas and implementation of parallelisation of the MOCO algorithm to reduce runtime is likely to yield additional improvements in fitness. The final full run of the algorithm for all 1842 electoral divisions took approximately 5 hours and 10 minutes.

The resulting synthetic population allows for individual risk factors to be aggregated and displayed visually at the individual level (Figs. 13.10 and 13.11) as well as an the individual student level (Fig. 13.12) and also allows for hotspot analysis which transcends geographic boundaries and can assist in defining more rational policy responses for tackling rural educational risk across the country. Further work

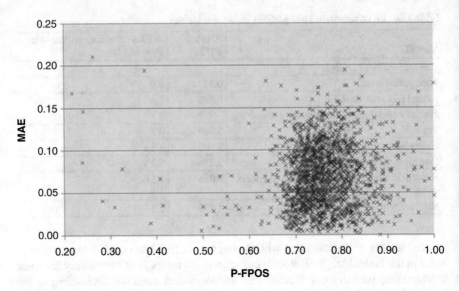

Fig. 13.8 Scatter plot of objective function values for all EDs in the synthetic population

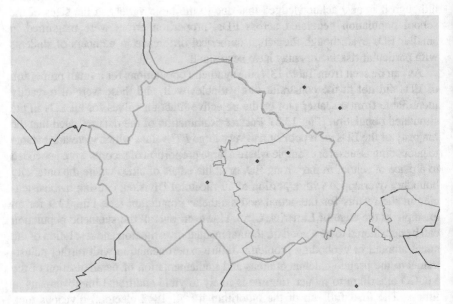

Fig. 13.9 City boundaries overlapping with ED boundaries. Notes: ED 21051 (highlighted with blue border) is overlapped by the edge of Limerick City (blue shading). As schools in cities were excluded from the analysis, this meant that a synthetic population was only generated for the school marked in green, which was compared against the marginal totals for the ED as a whole, causing large values of FT^2

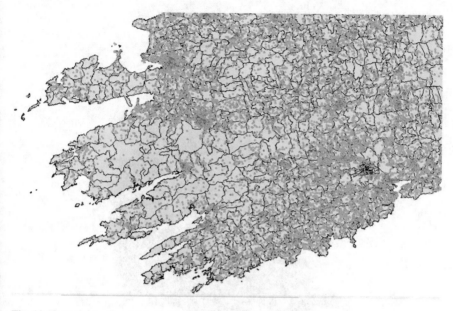

Fig. 13.10 Individual level risk profiles – South–West of Ireland

Fig. 13.11 Individual level risk profiles – Donegal

Fig. 13.12 School level relative risk profiles in Ireland aggregated from synthetic population

will develop this synthetic population into an agent-based model to study possible effects of interactions between members of this base population within both communities and schools, and how these interactions might amplify or attenuate risk, by prescribing a model derived from empirical results in this area. The model can also be used to test how different school structures in rural areas (for example, amalgamation of schools) or different types of policy interventions might nudge the system towards being structurally more conducive to promoting equity and reducing educational risk in the rural school population. Thus this work serves as a real-world example of how synthetic populations could be build and applied to assist in understanding and solving key challenges facing education policymakers.

References

Abowd, J. M., & Lane, J. (2004). New approaches to confidentiality protection: Synthetic data, remote access and research data centers. In *International workshop on privacy in statistical databases* (pp. 282–289). Berlin, Heidelberg: Springer.
Ainscow, M. (1999). *Understanding the development of inclusive schools*. London: Falmer Press.

Ballas, D., Clarke, G., Dorling, D., Eyre, H., Thomas, B., & Rossiter, D. (2005a). SimBritain: A spatial microsimulation approach to population dynamics. *Population, Space and Place, 11*(1), 13–34.

Ballas, D., Rossiter, D., Thomas, B., Clarke, G. P., & Dorling, D. (2005b). *Geography matters: Simulating the local impacts of national social policies.* York: Joseph Rowntree Foundation.

Boldt, S., Devine, B., McDevitt, D., & Morgan, M. (1998). *Educational disadvantage and early school leaving (demonstration programme on educational disadvantage).* Dublin: Combat Poverty Agency.

CSO. (2013). *Census of population of Ireland 2011 Place of Work, School or College Census of Anonymised Records (POWSCAR) user guide.* Central Statistics Office. http://www.cso.ie/en/census/census2011placeofworkschoolorcollege-censusofanonymisedrecordspowscar/

Czyzak, P., & Jaskiewicz, A. (1998). Pareto simulated annealing—A metaheuristic tech- nique for multiple-objective combinatorial optimization. *Journal of Multi-Criteria Decision Analysis, 7,* 34–47 12.

Department of Education and Skills. (2017). *DEIS plan 2017.* Department of Education and Skills. https://www.education.ie/en/Publications/Policy-Reports/DEIS-Plan-2017.pdf

Duh, J. D., & Brown, D. G. (2007). Knowledge-informed Pareto simulated annealing for multi-objective spatial allocation. *Computers, Environment and Urban Systems, 31*(3), 253–281.

Education Act 1998, IRL. http://www.irishstatutebook.ie/eli/1998/act/51/enacted/en/html

Ehrgott, M., & Gandibleux, X. (2004). Approximative solution methods for multiobjective combi-natorial optimization. *TOP, 12*(1), 1–63.

Galster, G. C., & Killen, S. P. (1995). The geography of metropolitan opportunity: A reconnais-sance and conceptual framework. *Housing Policy Debate, 6*(1), 7–43.

Goldstone, R. L. (2006). The complex systems see-change in education. *The Journal of the Learning Sciences, 15*(1), 35–43.

Green, T. L. (2015). Places of inequality, places of possibility: Mapping "opportunity in geogra-phy" across urban school-communities. *The Urban Review, 47*(4), 717–741.

Haase, T. (1994). *The identification of schools and pupils as disadvantaged: A preliminary assess-ment.* Dublin: Educational Research Centre.

Haase, T., McKeown, K., & Rourke, S. (1995). *Local development strategies for disadvantaged areas: Evaluationof the global grant, 1992–1995.* Dublin: Area Development Management Ltd.

Harding, A. (2007, August). *Challenges and opportunities of dynamic microsimulation modelling.* Plenary paper presented to the 1st General Conference of the International Microsimulation Association, Vienna (Vol. 21).

Harland, K., Heppenstall, A., Smith, D., & Birkin, M. (2012). Creating realistic synthetic popu-lations at varying spatial scales: A comparative critique of population synthesis techniques. *Journal of Artifical Societies and Social Simulation, 15*(1), 1–15.

Hauke, J., Lorscheid, I., & Meyer, M. (2017). Recent development of social simulation as reflected in JASSS between 2008 and 2014: A citation and co-citation analysis. *Journal of Artificial Societies and Social Simulation, 20*(1), 5. http://jasss.soc.surrey.ac.uk/20/1/5.html. https://doi.org/10.18564/jasss.3238.

Heckman, J. J. (2008). Schools, skills, and synapses. *Economic Inquiry, 46*(3), 289–324.

Helbing, D., & Balietti, S. (2011). From social data mining to forecasting socio-economic crises. *The European Physical Journal-Special Topics, 195*(1), 3–68.

Huynh, N. N., Barthelemy, J., & Perez, P. (2016). A heuristic combinatorial optimisation approach to synthesising a population for agent-based modelling purposes. *Journal of Artificial Societies and Social Simulation, 19*(4), 11.

Hynes, S., Morrissey, K., O'Donoghue, C., & Clarke, G. (2009). A spatial micro-simulation analy-sis of methane emissions from Irish agriculture. *Ecological Complexity, 6*(2), 135–146.

Innes, J. E., & Booher, D. E. (1999). Consensus building and complex adaptive systems: A frame-work for evaluating collaborative planning. *Journal of the American Planning Association, 65*(4), 412–423.

Kellaghan, T. (1995). *Educational disadvantage in Ireland*. (No. 20). Dublin: Combat Poverty Agency.

Macal, C. M., & North, M. J. (2010). Tutorial on agent-based modelling and simulation. *Journal of Simulation, 4*(3), 151–162.

Meyer, M., Lorscheid, I., & Troitzsch, K. G. (2009). The development of social simulation as reflected in the first ten years of JASSS: A citation and co-citation analysis. *Journal of Artificial Societies and Social Simulation, 12*(4), 12.

Nakaya, T., Fotheringham, A. S., Hanaoka, K., Clarke, G., Ballas, D., & Yano, K. (2007). Combining microsimulation and spatial interaction models for retail location analysis. *Journal of Geographical Systems, 9*(4), 345–369.

OECD. (2016). *New approaches to economic challenges (NAEC), insights into complexity and policy*. Paris: OECD Publishing.

Parsons, E., Chalkley, B., & Jones, A. (2000). School catchments and pupil movements: A case study in parental choice. *Educational Studies, 26*(1), 33–48.

Pearce, J. (2000). Techniques for defining school catchment areas for comparison with census data. *Computers, Environment and Urban Systems, 24*(4), 283–303.

Reiter, J. P., & Mitra, R. (2009). Estimating risks of identification disclosure in partially synthetic data. *Journal of Privacy and Confidentiality, 1*(1), 6.

Rich, J., & Mulalic, I. (2012). Generating synthetic baseline populations from register data. *Transportation Research Part A: Policy and Practice, 46*(3), 467–479.

Ruggles, S. (2014). Big microdata for population research. *Demography, 51*(1), 287–297.

Tate, I. V., & William, F. (2008). "Geography of opportunity": Poverty, place, and educational outcomes. *Educational Researcher, 37*(7), 397–411.

Voas, D., & Williamson, P. (2000). An evaluation of the combinatorial optimisation approach to the creation of synthetic microdata. *Population, Space and Place, 6*(5), 349–366.

Voas, D., & Williamson, P. (2001). Evaluating goodness-of-fit measures for synthetic microdata. *Geographical and Environmental Modelling, 5*(2), 177–200.

Weir, S., Archer, P., & Millar, D. (2009). *Educational disadvantage in primary schools in rural areas*. Dublin: Educational Research Centre.

Wheaton, W. D., Cajka, J. C., Chasteen, B. M., Wagener, D. K., Cooley, P. C., Ganapathi, L., et al. (2009). Synthesized population databases: A US geospatial database for agent-based models. *Methods Report (RTI Press), 2009*(10), 905.

Chapter 14
Modelling Collaborative Knowledge Creation Processes: An Empirical Application to the Semiconductor Industry

Martina Neuländtner, Manfred Paier, and Astrid Unger

Introduction

Being the fundamental basis for generating new innovations, the creation of new knowledge constitutes a crucial factor for organizations to be competitive, in particular in times of converging technologies and increasing market pressures due to more rapidly changing patterns of demand in a globalizing world (see e.g. Bathelt et al. 2004). More and more frequently, the creation of knowledge is the result of interactive, collaborative learning processes among organizations of different types located in different geographical spaces; especially, in a strongly knowledge-based economy, collaborative knowledge creation is increasingly gaining importance (Hoekman et al. 2010).

In the process of collaborative knowledge creation, networks of R&D (Research & Development) relationships between firms, universities and research organizations constitute essential means by which knowledge flows between these actors, and by this, enable access to external, new sources of knowledge. While such knowledge flows are considered to be mostly geographically localized within regions or nations due to its 'sticky' nature, R&D networks are assumed to serve as channels for transmitting knowledge over larger geographical distances (see e.g. Autant-Bernard et al. 2007). Thus, attention has recently been shifted to the investigation and modelling of regional knowledge creation processes by means of their region-internal and region-external knowledge interactions in the form of R&D collaborations (e.g. Cowan et al. 2006; Scherngell 2013 for recent overview).

Up to now, we can find a number of empirical studies that have investigated the structure and dynamics of R&D networks, and its role for knowledge creation and innovation in the widest sense, using spatial econometrics (e.g. Hoekman et al.

M. Neuländtner (✉) · M. Paier · A. Unger
AIT Austrian Institute of Technology, Center for Innovation Systems & Policy, Vienna, Austria
e-mail: martina.neulaendtner@ait.ac.at; manfred.paier@ait.ac.at; astrid.unger@ait.ac.at

© Springer Nature Switzerland AG 2019
D. Payne et al. (eds.), *Social Simulation for a Digital Society*, Springer Proceedings in Complexity, https://doi.org/10.1007/978-3-030-30298-6_14

2010; Wanzenböck and Piribauer 2016) and Social Network Analysis (SNA; e.g. Glückler 2007; Maggioni and Uberti 2009), or combinations of them – both, at the organizational level (e.g. Ahuja 2000; Zaheer and Bell 2005; Giuliani 2007), and from a spatial perspective (e.g. Whittington et al. 2009; Maggioni and Uberti 2011).

In addition, more and more simulation studies emerge in the literature in this vein. However, mostly of theoretical and stylized nature; for instance, simulation models on the emergence of innovation networks (Savin and Egbetokun 2016; Mueller et al. 2017), the diffusion of knowledge and innovations within networks (Alkemade and Castaldi 2005; Cowan and Jonard 2003), or both (Deroıan 2002; März et al. 2006). Simulation studies in this respect, with aspiration to empirical representability by employing empirical data or empirical calibration and validation techniques are scarce and limited to certain fields of application such as biotechnology (Paier et al. 2017), agriculture (Berger 2001), environment (Schwarz and Ernst 2009), or the financial market (LeBaron 2001).

In this study, we propose an empirical agent-based model of knowledge creation in a system of interacting R&D actors that complements these existing approaches by accounting for micro-level dynamics, such as heterogenous research and collaboration processes.

In an illustrative application to the Austrian semiconductor industry, we pursue two objectives: (i) modelling knowledge creation processes in a characteristic knowledge-driven industry with a special emphasis on R&D collaborations, and (ii) investigating the impact of R&D networks (measured in terms of collaboration intensity, centralization, and density) on knowledge production output at the country level (measured by means of patents), especially investigating the role of international collaborations for knowledge creation.

By this, the study proposes an interesting combination of two methodological entry points, rather new to the literature so far: (i) Agent-based Modelling (ABM) that provides a framework to simulate behavior and interactions of heterogeneous agents within a given environment, with the objective to model the complexity of real-world systems (Nikolic et al. 2013) and (ii) Social Network Analysis (SNA), which provides valuable tools to characterize the local embeddedness of organizations in R&D networks and their global structures and dynamics. SNA is generally used to analyze the connectivity structure between individual actors or organizations (e.g. Ter Wal and Boschma 2009; Graf 2011).

Therefore, we contribute to the state of the art in two major respects: *First*, we combine ABM and SNA and hence, make use of the advantages of each approach in order to address the research objectives in a comprehensive manner. And *second*, we accompany this unique mix of methods with a sound and elaborate use of qualitative and quantitative data for the initialization, calibration and validation of the model that enables us to apply our model to real world contexts.

The remainder of this chapter is organized as follows. First, we outline recent perspectives on collaborative knowledge creation processes and the importance of inter-organizational R&D networks along with their effects on knowledge creation. Then, we characterize the specifics of collaborative knowledge creation processes with respect to the semiconductor industry. The subsequent section presents the agent-based simulation model emphasizing the complexity of collaborative research processes. The next part comprises the empirical framework, including the initialization as well as the calibration and validation of the model. And then the results of the scenario analyses – targeting the effects of R&D networks, and internationalization on knowledge creation output – are presented. Finally, we highlight the main findings and draw conclusions for further research.

The Networked Nature of Collaborative Knowledge Creation

Knowledge creation is understood as the formation of new ideas and is generally considered an interactive collaborative learning process. Knowledge in general subsumes organized or structured information that is difficult to codify and interpret, often due to its intrinsic complexity and indivisibility (Karlsson and Gråsjö 2014). In principle, new knowledge can be created within an organization by means of internal research and development (R&D) but also at an inter-organizational level relying on informal and formal interactions, ranging from simple network activities of researchers to long-term and contract-based arrangements. Especially, inter-organizational collaborations are considered as an indispensable and increasingly important element for an organization's knowledge creation[1].

It is widely acknowledged that an organization's R&D performance on the one hand, depends on certain organizational characteristics, such as a firm's absorptive capacity (Cohen and Levinthal 1990; Rothaermel and Alexandre 2009), its R&D intensity (Laursen and Salter 2004), size (Belderbos et al. 2004; Negassi 2004), industry affiliation (Veugelers and Cassiman 2005), and previous involvement in collaborative R&D (Laursen and Salter 2004; Capaldo 2007; Paier and Scherngell 2011). On the other hand, the network structure and an organization's network position have considerable impact on knowledge creation and beyond that, the innovative performance. For instance, studies on the effect of an actor's embeddedness in a knowledge network on its innovative performance are manifold and exist for different industries such as biotechnology (Powell et al. 1996; Salman and Saives 2005), chemicals (Ahuja 2000; Gilsing et al. 2008) or banking (Zaheer and Bell 2005). Most of the studies find evidence for a positive relationship between network embeddedness and innovativeness (e.g. Ahuja 2000; Salman and Saives 2005; Ting Helena Chiu 2008). However, Uzzi (1997) reveals a paradox of embeddedness indicating difficulties for organizations to access new opportunities and hence, change their network portfolio as they become too embedded in network relationships, since isomorphism within the network decreases diversity (see also Bathelt and Taylor 2002; Sofer and Schnell 2002).

---

[1] However, it is argued that inter-organizational network channels are by no means sufficient but rather considered complementary to internal capabilities, since similar internal capabilities are necessary to evaluate research done by collaboration partners (e.g. Inkpen and Tsang 2005; Cowan and Jonard 2009).

In particular, the spatial dimension of the individual actor's embeddedness in networks of knowledge creation gained attention in regional economics and economic geography, since collaborations also act as a driver for innovation and economic growth in regions and countries (Romer 1990). In this regard, the debate is driven by the notion of "local buzz" and "global pipelines" as two forms of collaborative knowledge creation (Bathelt et al. 2004). Whereas, "local buzz" highlights the facilitative role of proximity for knowledge creation, the idea of "global pipelines" point to the necessity of global knowledge creation and search processes across regional boundaries. As for the individual actors, the inward flow of novel ideas, knowledge and skills may reduce the risk of technological and cognitive lock-in by creating new impulses. However, simply aiming for high outward orientation (large number of external network links) could also have negative consequences for regions, e.g. by neglecting the region-internal capabilities (Broekel et al. 2015).

Moreover, many scholars address the effects of structural holes in an actor's knowledge network on the ability to create new knowledge (Ahuja 2000; Zaheer and Bell 2005; Vasudeva et al. 2013). In general, a globally dense knowledge network (i.e. a well-connected network) is considered to facilitate the creation of strong ties and hence, and is assumed to promote knowledge diffusion, also at the regional level. Nevertheless, a high network density can also be considered to somehow limit the potential for novelty creation, since redundant ties reduce the flow of new and unique ideas (Ahuja 2000; Gilsing et al. 2008; Molina-Morales and Martínez-Fernández 2010).

R&D Collaborations in the Semiconductor Industry

Semiconductors are considered key technologies with strategic importance for a broad range of industries, regions and whole economies (OECD 2004; European Commission 2006a, b). They are characterized by rapid technological change with short product life cycles and a correspondingly high potential for innovation. Significant features of the semiconductor industry itself are global R&D and production networks and high capital intensity. In this situation, it is particularly important for companies to reduce uncertainty regarding their technology strategy by engaging in international research. For instance, this is targeted through the development of industry-wide roadmaps (Arden et al. 2011).

Hence, steady R&D activities and particularly R&D collaborations constitute central elements in the corporate strategies of semiconductor firms. In general, R&D collaborations in the semiconductor industry are characterized by a value chain subsuming vertical partners such as suppliers and customers as well as competitors as horizontal partners (Belderbos et al. 2015). R&D collaborations may span parts or the whole value chain but may also occur within one segment of the value chain. However, in that respect, competition is an issue not to be neglected.

Along an industry's value chain, firms differ with respect to their field of technology, orientation regarding their innovation and collaboration, as well as their patenting strategy. Moreover, knowledge and research intensity are not evenly distributed along the value chain and across the different technology segments; hence, following specified innovation and collaboration strategies is essential for some firms, whereas it is not for others. Generally, mostly vertical collaborations are characterized by persistence; especially, relatively localized R&D relations between larger semiconductor firms and smaller supplier firms. These collaborations constitute strong ties, based on trust.

Especially, for smaller regions and nations, where knowledge is concentrated in space and rather specialized, since semiconductor firms generally do not cover the whole value chain, and mostly only encompass a rather small range of technologies, embeddedness in international R&D networks is of great importance. Knowledge created or accessed by means of international relations may be incorporated into intra-regional knowledge diffusion mechanisms and regional knowledge bases (e.g. Bathelt et al. 2004). By tapping into external knowledge, regional disparities regarding the innovatory potential may be reduced and a lock-in could be prevented (see e.g. Boschma and Ter Wal 2007). However, respective technological expertise must be present within organizations, regions and nations to efficiently make use of the newly accessed and created knowledge.

Overview of the Agent-Based Model

The underlying conceptual model is derived from theoretical considerations and empirical observations in innovation economics, intending to conduct scenario simulations applicable to real-world contexts. We start from a well-known concept of knowledge representation in agent-based modelling, the SKIN model (Simulating Knowledge dynamics in Innovation Networks), a multi-agent model containing heterogenous agents which act in a complex and changing environment (Gilbert et al. 2001; Pyka et al. 2002; Ahrweiler et al. 2004) and follow a recent guideline for empirical ABMs (Smajgl and Barreteau 2014). Core aspects of this approach are to use qualitative data (expert knowledge) and quantitative data (patent and company databases, statistical information) in the design, calibration and validation of the model.

An overview of elements of the simulation model is given in Fig. 14.1. In essence, the proposed model of knowledge creation builds upon recent work by Paier et al. (2017) that is refined and specified to resemble the semiconductor industry. The model comprises three elements: (i) heterogeneous agents, (ii) their relations and activities, and (iii) the agents' environment.

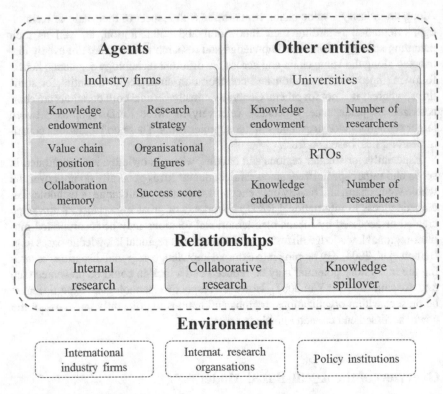

Fig. 14.1 Model architecture

Agents and Attributes

Agents differ with respect to their location (national or international), organization type (industry firm, university or public research institute) and form of organization (private or public). Related to these dimensions they exhibit a different set of attributes and strategies. The relations and activities of the agents represent the (collaborative) research processes aiming at the creation of knowledge. The agents, relationships and activities are embedded in an environment representing the research system with sector-specific institutional characteristics. As an external element, the environment subsumes international agents and policy institutions.

National Industry Firms[2] are specified with regard to their knowledge endowment, research strategies, value chain position, organizational figures, collaboration memory and success score.

---

[2] It is important to note that the model processes, if not otherwise stated, refer to the national industry firm population, since only they are qualified to actively perform research (with the possibility to engage with the other agent populations – international industry firms, national/international research universities and research institutes).

Knowledge Endowment. The knowledge endowment of an agent is defined as a set of so-called kenes (Gilbert et al. 2001). How many kenes the agent holds depends on its number of employees. Each kene consists of a technology class, a subfield associated with the respective technology class and an expertise level in the specific technology class and subfield. The technology classes are initialized according to the corresponding firm's patent portfolio. The subfield and the expertise level are random numbers between one and ten.

Research Strategies. The knowledge creation model foresees four different research strategies, that define how the agent chooses a new research target, i.e. *inactive, conservative, incremental* and *radical*. Each time step, an agent engages in research activities by defining a research target. For that purpose, the agent randomly chooses one of the three existing kenes and modifies it according to its given research strategy. Dependent on the research success, the chosen strategy may be changed.

Value Chain Position. The model's value chain comprises four basic elements – suppliers, semiconductors, microsystems and customers, that each is divided in further subgroups. Each agent is, based on expert knowledge, manually assigned to one position in the value chain. Dependent on the position held, different collaboration probabilities with respect to collaboration partners along the value chain apply.

Organizational Figures. To further introduce heterogeneity into the agent population each agent is individually characterized by organizational figures, namely research expenditures, number of employees, assets and age. These figures, representing the fitness of an agent, are essential for the evaluation of the agent's research success in the output part (i.e. patent evaluation).

Collaboration Memory. In the model, the persistence of collaboration partnerships is implemented by means of a collaboration memory of the agents, meaning that once having performed successful collaborative research, the partner is remembered and is, subsequently, in a preferential position of being chosen again. Obviously, former partners can also be forgotten, if there were no successful interactions over a longer period of time.[3]

Success Score. The agents' success score is a counter of successful and failed research efforts. Whenever an agent successfully performed research (i.e. the research target set was found), the counter is increased by one and, conversely, decreased by one in the case of unsuccessful research. The success score is used as

---

[3] The collaboration memory is implemented by means of an ordered list of length ten, representing equally preferred collaboration partners. The most recent collaboration partner is put in the first position in the list. Hence, with each new collaboration a former partner is pushed back on the next position in the list – eventually, dropping out of the collaboration memory after – at least ten – ticks, representing ten quarters of a year.

a performance indicator and hence, serves the purpose to determine changes in the research strategy of the agent.

Universities and Research and Technology Organizations (RTOs) take the position of passive knowledge suppliers during collaborative research[4]. Hence, they are themselves not equipped with research strategies but rather exhibit a knowledge portfolio, potential collaborative partners can access. They are characterized by a knowledge portfolio, and number of researchers.

Relations and Activities

In order to perform research and innovate, firms may employ different knowledge acquisition strategies (see Fig. 14.2): One way to gain knowledge is to create it within the organization, i.e. *internal research*. This requires certain internal capabilities, such as a sufficient knowledge stock, skilled researchers and an appropriate institutional setting. In addition to internal research, new knowledge can be created by means of inter-organizational collaborations, i.e. *collaborative research*. Engaging with other firms in partnerships or networks provides access to knowledge and resources that are otherwise unavailable (Powell et al. 1996). A third option to acquire knowledge is through *knowledge spillover*. Based on the assumption that a certain amount of knowledge is non-excludable and a non-rival good (Fischer and Fröhlich 2001), knowledge can spill over from one agent to another, for instance due to informal network interactions. The attainment of the research result depends on respective success rate parameters. If the research result is actually achieved through these research processes, the new kene replaces the old one in the knowledge endowment of the agent.

Whether the knowledge gains of an agent classify for becoming a patent, is determined by an empirical output filter (*fitness function*). The fitness function is composed of two parts: a system parameter and a function including the empirically estimated coefficients influencing the patenting propensity of an individual agent given its respective organizational figures. The coefficients are estimated by means of a Poisson regression model (see Cameron and Trivedi 2013).

---

[4] Research organisations, such as universities, are – in contrast to firms – dedicated to R&D and usually exhibit a broad variety of topics and personnel for research projects, making it easy for firms to find suitable collaboration partners. Moreover, the focus of the model is to simulate collaborations by firms dependent on their research strategies; modelling in detail public knowledge infrastructure, such as universities, would go beyond the scope of this simulation model.

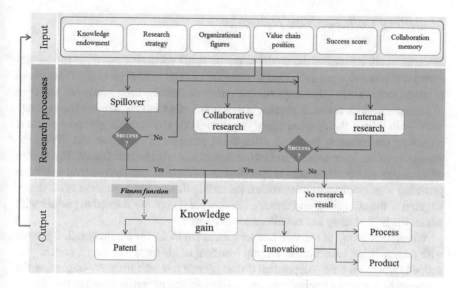

Fig. 14.2 Agents' attributes and processes

Environment

The agents' environment subsumes international agents that are not part of the national innovation system, as well as policy institutions.

International Agents International research links due to collaboration, be it be with suppliers, customers, competitors, within the own corporate group or with universities and research institutes, is seen as a crucial factor to gain access to new knowledge and markets. Hence, it is essential to ensure one's competitiveness in the future. However, it has to be noted that knowledge from geographically distant contexts may be difficult to understand, transfer and use (Capaldo and Messeni Petruzzelli 2015). In the simulation model, international industry firms, universities and research institutions are included. They, like the national universities and research institutes, serve as passive knowledge suppliers and are also equipped with the same attributes, namely a knowledge portfolio and the number of researchers.

Policy Institutions An innovation system includes not only companies and research organizations, but also institutional factors, such as public research funding, its design, organization and implementation. In the model, the main focus is not on the institutions themselves, but on the interventions, they take. These policy interventions, such as different R&D funding measures, enter the model by means of scenarios, exogenously determined by the modeler through variations of initial parameters.

Modelling Collaboration Networks

Since an explicit focus of the model is put on R&D collaboration, these model elements and processes are described in more detail. The formation of R&D networks is linked to the knowledge acquisition strategy of *collaborative research*. The core of the collaborative research process is the choice of a suitable partner to collaborate with. With whom the agent engages depends (i) on the suitable group of partners regarding the location, type and value chain position of the agent and (ii) on the proximity of the knowledge bases of the agents that wish to collaborate. The choice of a suitable group of partners is specified as a decision tree with several junctions provided with empirical probabilities. Preceding this selection process is the query targeted at the collaboration memory, checking if it contains a suitable partner with whom the firm already has collaborated in the past.

Within this group of suitable partners, an agent looks for a collaboration partner with a knowledge base similar, but not identical to his own. Therefore, certain predefined thresholds hold – dependent on the agent's research strategy – regarding the elements of the kenes (technology class, subfield and expertise level) in the knowledge endowment of the agent.

Underlying this process of partner choice are three distinct empirically based proximity networks: networks including information (i) on the location and organization type, (ii) on the value chain position and (iii) on the proximity of technology fields in order to determine the similarity of the agents' knowledge bases[5]. Together, they determine the probabilities of agents to collaborate within these networks, dependent on their own network position.

To reflect limitations in research capacity and budget restrictions, the frequency of potential research collaborations depends on the size of the agents (measured in terms of the number of employees), whereas, small agents may only engage in a research activity every six time steps (representing 6 months), medium sized and large firms may start new research collaborations every three time steps and every time step, respectively.

Empirical Underpinning

Empirical Initialization

The model is initialized at the agent level as well as at the system level. For the initialization at the agent level, we include 138 private Austrian firms from the semiconductor industry, 50 international private firms, and 39 Austrian and foreign

---

[5]To measure technological proximity between technology classes, we employ the Jaccard index based on the number of co-references of technology classes on patent documents that are derived from the empirical patent stock of the firm sample.

universities and public Research and Technology Organizations (RTOs). The selection of the agents was made based on their sector-specific patenting behavior during the period of 1990–2013.

The set of technology classes individually assigned to each agent corresponds to the patent classes in which they actually hold patents in the initialization period (2000–2008). The patent classes are used on a three-digit subclass level specified by the International Patent Classification (IPC). The number of technology classes in the model results from the most frequently occurring IPC classes on the patents of the firm sample (i.e. 73 technology classes).

For the initialization of an agent's knowledge endowment, first, for every kene a technology class is randomly drawn from the set of references to IPC classes (from the associated firm's individual patent stock). Second and third, the respective subfield and expertise level are chosen randomly. Moreover, each agent is also individually equipped with four empirically based organizational figures: research expenditure, number of employees, assets, and age, taken from the Orbis company database.

Empirical Calibration and Validation

Unlike the agent characteristics, which are initialized with empirical firm data, a set of free system parameters governs the model processes. To calibrate the model, these parameters are adjusted in such a way that the simulation output renders the empirical data of the real-world reference system.

Core of the empirical calibration is the fitting of model parameters in a way that the resulting output variables best fit or lie within a range of the selected empirical measures; calibration criteria are then defined accordingly. Thiele et al. (2014) point out two different strategies for fitting model parameters to observational data: (i) best-fit and (ii) categorical calibration.

Whereas, best-fit calibration is aimed at finding the parameter combination that best fits the observational data, using categorical calibration, a range of plausible values is defined for each calibration criterion. As proposed by Thiele et al. (2014) a hybrid approach by transforming the categorical criteria to a best-fit criterion is followed here. This is done by means of conditional equations and the definition of a cost function, evaluating the cost for a parameter value of not being in the acceptable value range. For each selected empirical measure, an acceptable value range is defined. If the simulated value lies within this interval, no costs are incurred. Otherwise, a cost factor based on the squared relative deviation to the mean value of the acceptable range is assigned. The final cost function is then defined as the sum of the individual costs of each criterion. Finally, the parameter combination with the lowest cost is chosen as the one that best fits the real-world system.

Empirically, two patent-related quantities are chosen as criteria for the cost function: (i) the total number of patents in the firm population and (ii) the patenting profile of the population, i.e. the distribution of these patents over the patent classes.

The model was empirically validated with industry experts during and at the end of the model conceptualization phase. Such expert validation is a commonly used approach to assure a close-to-reality depiction of industry processes, including the behavior of agents in the simulation model, and also its applicability to the specific purpose (Nikolic et al. 2013).

About 20 corporate R&D and innovation experts from a representative sample of Austrian semiconductor firms, as well as policy-makers were asked for their willingness to conduct a telephone interview based on a guideline provided in advance. Five experts agreed on an interview; five other experts agreed on informal discussions. In this way, information on sector characteristics, patenting strategies and activities as well as the role of Research, Technology and Innovation (RTI)-policy, was collected to lay down the foundation for the model conceptualization. Within the conceptualization phase, a group discussion among the initially interviewed experts, plus two additional industry experts was conducted where model assumptions and mechanisms were discussed subject to their plausibility. Based on this discussion, some adaptions to the model were made; for instance, an improved representation of long-term national collaborations based on trust (reflected by a collaboration memory) together with a strong focus on international R&D (reflected by including international firms, universities and research institutes), as well as an elaborated sector-specific value chain determining the collaboration behavior, became important pillars of the model.

Scenario Analyses and Results

As outlined in section "The Networked Nature of Collaborative Knowledge Creation", evidence on the effects of inter-firm networks on economic and innovative performance is manifold and exists for various sectors and industries. Here, we focus on the aspect of knowledge creation in the semiconductor industry and apply the model to analyze scenarios characterized by different collaborative behavior of the firms. The ABM approach allows to take into account how the structure of the collaboration network co-evolves with the heterogeneity of the actors as well as their embedding in technology space.

As a measure of knowledge output, we employ patents, which are generally considered a suitable indicator for knowledge creation, invention and innovation (Mansfield 1986) – which are closely linked – and are widely used by scholars in innovation research (e.g. Singh 2005; Gomes-Casseres et al. 2006). More specifically, the semiconductor industry is regarded as particulary patenting intensive (Wajsman et al. 2013). Nevertheless, the simulation model also includes product as well as process innovations as output indicators; however, since all output measures show high correlations, we present results regarding patents only, due to their empirical underpinning in the model.

Hence, in a first scenario analysis, we aim at investigating the *effects of increased R&D networking on knowledge creation* in the Austrian semiconductor industry, which is characterized by a strongly localized and specialized knowledge portfolio. Especially, to a certain degree a backbone structure is predominant, comprising local suppliers being linked to a few comparatively large and dominating firms. However, overall, the regional and national hubs strive for embeddedness in transnational, global R&D networks. Thus, in a second simulation experiment, we furthermore examine the *effect of increasing internationalization on knowledge created* by the national firm agents. The specifics of the Austrian semiconductor industry make it an interesting application promising novel insights that can – to a certain extent – be generalized to comparatively small and spatially concentrated, but highly vivid and specialized industries or economies. Both scenarios depart from the reference case (defined by the empirically calibrated parameter values of the model as described in section "Empirical Calibration and Validation") in terms of the collaborative behavior. Since these scenario analyses reveal antagonistic effects on knowledge creation, we finally inquire a third scenario, namely a *combination of intensifying collaboration in general and specifically increasing international collaboration.*

The model is implemented in Java using the simulation library MASON[6]. The simulations are conducted over 120 time steps, which is calibrated to correspond to a period of 10 years, and the results represent averages of 100 single runs with varying random seeds.

Effects of R&D Networks on Knowledge Creation

The first scenario targets the analysis of the effects of R&D networks on knowledge creation. We examine the effects of increasing willingness of the agents to cooperate in R&D with suitable other agents in general. The resulting collaboration networks are characterized with respect to three measures, allowing us to compare the different networks: (i) the number of collaboration links – representing the collaboration intensity, (ii) network density – describing how well connected the actors are being defined as the ratio of actual to potential links in the network, and (iii) the degree centralization – indicating a potential dominance structure of one or more agents in the network (see Wasserman and Faust 1994 for detailed description of network measures).

The scenario is constructed by stepwise increasing the share of (potentially) collaborative actors, which is implemented as an external system parameter (see Fig. 14.3). By this, evidently, the collaboration intensity is increased but also the centralization and density of the simulated R&D collaboration network are changed system inherently by endogenous processes in the model.

Increasing the collaboration intensity is a highly relevant scenario, closely linked to current policy initiatives to fund collaborative research on a national and European

---

[6] Upon request the code and detailed model documentation can be obtained from the authors.

level (e.g. EU Framework Programmes, EUREKA, JPI). What is striking at first glance, is the negative correlation between the total number of collaborations and the number of patents, which are used as a proxy for the knowledge created (see Fig. 14.3). Moreover, the degree centralization as well as the network density increase, as collaboration intensifies. This implies that higher values of network centralization and density generally relate to a lower number of patents.

These findings contradict evidence from other scholars suggesting a positive relationship between embeddedness, density and innovativeness of organizations (see section "The Networked Nature of Collaborative Knowledge Creation"). However, what becomes evident when looking at the model processes in detail, is the paradox of embeddedness as indicated by Uzzi (1997) emphasizing potential difficulties for organizations to access new opportunities. The model reflects these difficulties in the agent's choice of a suitable collaboration partner to perform research. Hereby, agents face restrictions in terms of (i) the range and quantity of knowledge (technology fields) being present in the national innovation system, (ii) the availability of partners with suitable knowledge portfolios in terms of knowledge similarity, and (iii) specifically for the semiconductor industry, being positioned in a value chain, meaning different probabilities to cooperate along this value chain.

Hence, the negative effect of an increased collaboration intensity on the knowledge output is largely determined by difficulties of national firm agents to find

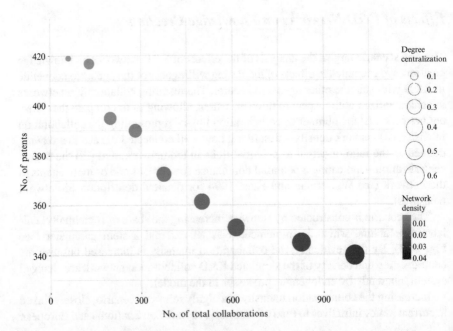

Fig. 14.3 Effects of R&D collaboration networks on knowledge output. Note: The nine cases in this scenario represent networks with increasing shares of collaborative agents, starting from 10% (leftmost) to 90% (rightmost) in steps of ten percentage points

suitable collaboration partners; not the fact that collaborations are unsuccessful. This highlights the importance of diversity in collaboration networks.

Taking a more detailed look at the negative relationship between network centralization, density and knowledge output, also suggests a certain lock-in due to the limited availability and restricted suitability of knowledge in order to engage in collaborative research activities. An increasing degree centralization along with higher collaboration intensity implies that few agents gain more and more dominance – as they are engaged in more collaborations than the others. Hence, the formation of new collaboration links is not very likely, although the number of potentially cooperative agents is being increased in this scenario setting; however, once agents have formed successful collaboration links, it is likely that they remain in these partnerships. Nevertheless, the increase in network density (i.e. the ratio of actual to potential links) suggests that still new collaboration links are formed.

Internationalization and Its Effects on Knowledge Creation

The Austrian semiconductor industry is characterized as a strongly localized and specialized industry with a small number of large international firms as its backbone. However, small national firms may face resource as well as structural barriers to engage in international collaborations and to benefit from global pipelines for the creation of new knowledge. With the second simulation experiment we investigate internationalization as a possibility to overcome this lock-in due to the structural limitations, i.e. with respect to the choice of suitable partners regarding the knowledge profile, value chain position, type of organization, and location.

This scenario is set up by gradually increasing the national firm agents' probability to collaborate preferably with international firms, universities and public research institutes, while the overall propensity to collaborate remains the same, i.e. at the empirical level. Hence, we address the impact of international collaborations on national knowledge output. We start from the empirically determined probability and create different levels of internationalization by increasing the probability of international collaborations step by step by five percentage points, respectively. By this, we obtain seven different levels, including the reference scenario based on the empirical probability.

Increasing the internationalization by increasing the probability of national agents to collaborate with international agents, has a clearly positive effect on the number of patents (see Fig. 14.4). The degree centralization as well as the network density exhibit only marginal changes. Whereas, high values of centralization coincide with low number of patents, there is obviously no distinct positive or negative impact of the network density on the knowledge output.

By specification of the scenario, the overall collaboration probability of the agents does not change; consequently, the positive correlation of the number of international collaborations and patent output arises from the increased tendency towards international collaborations. With international agents being empirically

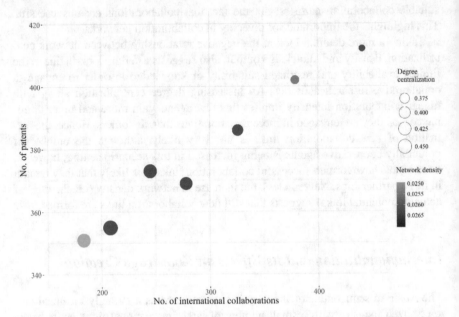

Fig. 14.4 Effects of internationalization on knowledge output. Note: The seven cases in this scenario represent networks with increasing shares of agents collaborating at an international level, starting from the empirical value of 59% (leftmost), and stepwise increasing shares by five percentage points, up to 89% (rightmost). Hereby, the overall share of collaborating agents is assumed to remain at 70%

equipped with heterogenous knowledge profiles spanning a wide variety of technologies, international collaboration links serve as vehicles for national agents to receive new knowledge input. The wide variety and quantity of knowledge available outside national borders facilitates the search of national agents for a suitable collaboration partner significantly – leading to more successful collaborations and beyond that, patents. Moreover, the newly accessed external knowledge is immediately incorporated in the national agents' own knowledge bases, which in turn allows them to find also new suitable national partners.

Increased R&D Collaboration and Internationalization

Now we turn attention towards a combination of the two scenarios, namely a scenario of increased internationalization and increased collaboration intensity. The simulation results show that while the number of patents decreases with an increased number of total collaborations, concurrent strengthening internationalization dampens this negative effect (see Fig. 14.5).

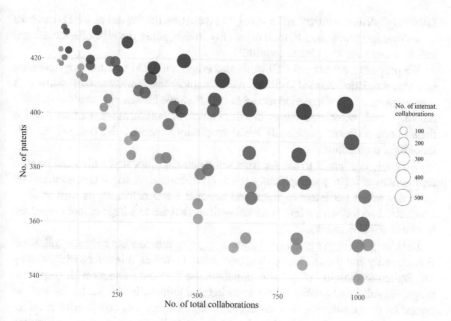

Fig. 14.5 Effects of increased collaboration intensity and internationalization. Note: The degree of shading corresponds to the levels of internationalization from the scenario analysis; i.e. the darker, the higher the internationalization. Hereby, the parameter values regarding collaboration and internationalization are taken from the preceding scenario simulations (as exhibited in Figs. 14.3 and 14.4)

The group with the lightest blue shading corresponds to the scenario with no additional internationalization and shows the presented decrease in patent output (section "Effects of R&D Networks on Knowledge Creation") together with a slight increase in international collaborations. As the share of internationally collaborating agents increases (darker bubbles) we observe a less negative relation between collaboration and patent output.

Summing up, the simulation results highlight that a simple increase in R&D collaboration propensity on the part of the firms does not necessarily lead to an improved knowledge creation performance. The crucial point – with respect to successful local knowledge creation in a global industry – is to have access to new knowledge from outside the local industrial ecosystem.

Concluding Remarks

This study has focused on modelling collaborative knowledge creation processes to analyze the impact of intra- and international R&D networks, in particular two research objectives were pursued: *first*, to model knowledge creation processes in a knowledge-driven industry and *second*, to investigate the impact of R&D networks on knowledge production output, especially investigating the role of international collaborations for knowledge creation.

We propose a robust and well calibrated empirical ABM of knowledge creation in a characteristic knowledge-driven sector, namely the semiconductor industry. A core element of the simulation model is collaborative knowledge creation in R&D networks and, associated therewith, the choice of collaboration partners, which draws upon different empirically based proximity networks, determining the collaboration probabilities.

We apply our model to the Austrian semiconductor industry, which can be seen representative for a geographically relatively localized sector – characterized by generally rather persistent regional and national R&D networks, as well as for a specialized sector, with a few dominating firms that are to a large extent embedded in global R&D networks.

Results of the scenario analyses show a negative relationship between collaboration intensity and the knowledge output created. However, this can be explained by the limited availability of suitable collaboration national partners with respect to range, quantity and quality of the knowledge dominant in this sector, as well as regarding the positioning in a value chain, representing e.g. competitive relationships and supplier networks. Also, the degree centralization as well as the network density exhibit a negative correlation with the knowledge output. Both findings indicate a lock-in resulting from the limited national knowledge sources.

Findings from the scenario of increased internationalization support the above conclusions. As the probability of national agents engaged in international collaborations increases (with constant overall collaboration probability), the number of patents, as an indicator for knowledge created, increases. Hence, we show that international collaborations have great potential to overcome such a lock-in situation by generating access to new and diverse knowledge that can be incorporated into the organizational and regional knowledge bases. However, this is only possible if the national R&D organizations themselves have respective expertise in key technology fields.

It is important to note that we explicitly focus on a model of regional knowledge creation and hence do not aim to explain any aspects of patent valuation and exploitation of inventions. However, future research will advance the model by approximating the economic value of knowledge created using patent and scientific citations.

References

Ahrweiler, P., Pyka, A., & Gilbert, N. (2004). Simulating knowledge dynamics in innovation networks (SKIN). In *Industry and labor dynamics: The agent-based computational economics approach world scientific* (pp. 284–296). Singapore: World Scientific Pub. Co..

Ahuja, G. (2000). Collaboration networks, structural holes, and innovation: A longitudinal study. *Administrative Science Quarterly, 45*(3), 425–455.

Alkemade, F., & Castaldi, C. (2005). Strategies for the diffusion of innovations on social networks. *Computational Economics, 25*(1–2), 3–23.

Arden, W., Brillouët, M., Cogez, P., Graef, M., Huizing, B., Mahnkopf, R., Pelka, J., Pfeiffer, J.-U., Rouzaud, A., Tartagni, M., Van Hoof, C., & Wagner, J. (2011). Towards a "More-than-Moore" roadmap. In *Report from the CATRENE Scientific Committee*. Paris, France: Cluster for Application and Technology Research in Europe on Nanoelectronics.

Autant-Bernard, C., Mairesse, J., & Massard, N. (2007). Spatial knowledge diffusion through collaborative networks. *Papers in Regional Science, 86*(3), 341–350.

Bathelt, H., Malmberg, A., & Maskell, P. (2004). Clusters and knowledge: Local buzz, global pipelines and the process of knowledge creation. *Progress in Human Geography, 28*(1), 31–56.

Bathelt, H., & Taylor, M. (2002). Clusters, power and place: Inequality and local growth in time–space. *Geografiska Annaler: Series B, Human Geography, 84*(2), 93–109.

Belderbos, R., Carree, M., & Lokshin, B. (2004). Cooperative R&D and firm performance. *Research Policy, 33*(10), 1477–1492.

Belderbos, R., Carree, M., Lokshin, B., & Sastre, J. F. (2015). Inter-temporal patterns of R&D collaboration and innovative performance. *The Journal of Technology Transfer, 40*(1), 123–137.

Berger, T. (2001). Agent-based spatial models applied to agriculture: A simulation tool for technology diffusion, resource use changes and policy analysis. *Agricultural Economics, 25*(2–3), 245–260.

Boschma, R. A., & Ter Wal, A. L. (2007). Knowledge networks and innovative performance in an industrial district: The case of a footwear district in the south of Italy. *Industry and Innovation, 14*(2), 177–199.

Broekel, T., Brenner, T., & Buerger, M. (2015). An investigation of the relation between cooperation intensity and the innovative success of German regions. *Spatial Economic Analysis, 10*(1), 52–78.

Cameron, A. C., & Trivedi, P. K. (2013). *Regression analysis of count data*. Cambridge: Cambridge University Press.

Capaldo, A. (2007). Network structure and innovation: The leveraging of a dual network as a distinctive relational capability. *Strategic Management Journal, 28*(6), 585–608.

Capaldo, A., & Messeni Petruzzelli, A. (2015). Origins of knowledge and innovation in R&D alliances: A contingency approach. *Technology Analysis & Strategic Management, 27*(4), 461–483.

Cohen, W. M., & Levinthal, D. A. (1990). Absorptive capacity: A new perspective on learning and innovation. *Administrative Science Quarterly, 35*, 128–152.

Cowan, R., & Jonard, N. (2003). The dynamics of collective invention. *Journal of Economic Behavior & Organization, 52*(4), 513–532.

Cowan, R., & Jonard, N. (2009). Knowledge portfolios and the organization of innovation networks. *Academy of Management Review, 34*(2), 320–342.

Cowan, R., Jonard, N., & Zimmermann, J.-B. (2006). Evolving networks of inventors. *Journal of Evolutionary Economics, 16*(1–2), 155–174.

Deroïan, F. (2002). Formation of social networks and diffusion of innovations. *Research Policy, 31*(5), 835–846.

European Commission. (2006a). Effects of ICT capital on economic growth. In *Innovation policy – Technology for innovation, ICT industries and E-business*. Brussels: European Commission.

European Commission. (2006b). Effects of ICT production on aggregate labour productivity growth. In *Innovation policy – Technology for innovation, ICT industries and E-business*. Brussels: European Commission.

Fischer, M. M., & Fröhlich, J. (2001). Knowledge, complexity and innovation systems: Prologue. In *Knowledge, complexity and innovation systems* (pp. 1–17). Berlin, Heidelberg: Springer.

Gilbert, N., Pyka, A., & Ahrweiler, P. (2001). Innovation networks – A simulation approach. *Journal of Artificial Societies and Social Simulation, 4*(3), 1–13.

Gilsing, V., Nooteboom, B., Vanhaverbeke, W., Duysters, G., & van den Oord, A. (2008). Network embeddedness and the exploration of novel technologies: Technological distance, betweenness centrality and density. *Research Policy, 37*(10), 1717–1731.

Giuliani, E. (2007). The selective nature of knowledge networks in clusters: Evidence from the wine industry. *Journal of Economic Geography, 7*(2), 139–168.

Glückler, J. (2007). Economic geography and the evolution of networks. *Journal of Economic Geography, 7*(5), 619–634.

Gomes-Casseres, B., Hagedoorn, J., & Jaffe, A. B. (2006). Do alliances promote knowledge flows? *Journal of Financial Economics, 80*(1), 5–33.

Graf, H. (2011). Gatekeepers in regional networks of innovators. *Cambridge Journal of Economics, 35*(1), 173–198.

Hoekman, J., Frenken, K., & Tijssen, R. J. (2010). Research collaboration at a distance: Changing spatial patterns of scientific collaboration within Europe. *Research Policy, 39*(5), 662–673.

Inkpen, A. C., & Tsang, E. W. (2005). Social capital, networks, and knowledge transfer. *Academy of Management Review, 30*(1), 146–165.

Karlsson, C., & Gråsjö, U. (2014). Knowledge flows, knowledge externalities, and regional economic development. In M. M. Fischer & P. Nijkamp (Eds.), *Handbook of regional science* (pp. 413–437). Berlin, Heidelberg: Springer.

Laursen, K., & Salter, A. (2004). Searching high and low: What types of firms use universities as a source of innovation? *Research Policy, 33*(8), 1201–1215.

LeBaron, B. (2001). Empirical regularities from interacting long-and short-memory investors in an agent-based stock market. *IEEE Transactions on Evolutionary Computation, 5*(5), 442–455.

Maggioni, M. A., & Uberti, T. E. (2009). Knowledge networks across Europe: Which distance matters? *The Annals of Regional Science, 43*(3), 691–720.

Maggioni, M. A., & Uberti, T. E. (2011). Networks and geography in the economics of knowledge flows. *Quality & Quantity, 45*(5), 1031–1051.

Mansfield, E. (1986). Patents and innovation: An empirical study. *Management Science, 32*(2), 173–181.

März, S., Friedrich-Nishio, M., & Grupp, H. (2006). Knowledge transfer in an innovation simulation model. *Technological Forecasting and Social Change, 73*(2), 138–152.

Molina-Morales, F. X., & Martínez-Fernández, M. T. (2010). Social networks: Effects of social capital on firm innovation. *Journal of Small Business Management, 48*(2), 258–279.

Mueller, M., Bogner, K., Buchmann, T., & Kudic, M. (2017). The effect of structural disparities on knowledge diffusion in networks: An agent-based simulation model. *Journal of Economic Interaction and Coordination, 12*(3), 613–634.

Negassi, S. (2004). R&D co-operation and innovation a microeconometric study on French firms. *Research Policy, 33*(3), 365–384.

Nikolic, I., Van Dam, K. H., & Kasmire, J. (2013). Practice. In K. H. Van Dam, I. Nikolic, & Z. Lukszo (Eds.), *Agent-based modelling of socio-technical systems* (Vol. 9, pp. 73–140). Dordrecht: Springer.

OECD. (2004). *The economic impact of ICT – Measurement, evidence and implications* (pp. 72–73). Paris: OECD.

Paier, M., Dünser, M., Scherngell, T., & Martin, S. (2017). Knowledge creation and research policy in science-based industries: An empirical agent-based model. In *Innovation networks for regional development* (pp. 153–183). Cham: Springer.

Paier, M., & Scherngell, T. (2011). Determinants of collaboration in European R&D networks: Empirical evidence from a discrete choice model. *Industry and Innovation, 18*(1), 89–104.

Powell, W. W., Koput, K. W., & Smith-Doerr, L. (1996). Interorganizational collaboration and the locus of innovation: Networks of learning in biotechnology. *Administrative Science Quarterly, 41*, 116–145.

Pyka, A., Gilbert, N., & Ahrweiler, P. (2002). Simulating innovation networks. In *Innovation networks: Theory and practice* (pp. 169–196). Cheltenham: Edward Elgar.

Romer, P. M. (1990). Endogenous technological change. *Journal of Political Economy, 98*(5, Part 2), S71–S102.

Rothaermel, F. T., & Alexandre, M. T. (2009). Ambidexterity in technology sourcing: The moderating role of absorptive capacity. *Organization Science, 20*(4), 759–780.

Salman, N., & Saives, A. L. (2005). Indirect networks: An intangible resource for biotechnology innovation. *R&D Management, 35*(2), 203–215.

Savin, I., & Egbetokun, A. (2016). Emergence of innovation networks from R&D cooperation with endogenous absorptive capacity. *Journal of Economic Dynamics and Control, 64*, 82–103.

Scherngell, T. (2013). *The geography of networks and R & D collaborations.* Berlin, Heidelberg/ New York: Springer.

Schwarz, N., & Ernst, A. (2009). Agent-based modeling of the diffusion of environmental innovations—An empirical approach. *Technological Forecasting and Social Change, 76*(4), 497–511.

Singh, J. (2005). Collaborative networks as determinants of knowledge diffusion patterns. *Management Science, 51*(5), 756–770.

Smajgl, A., & Barreteau, O. (2014). *Empirical agent-based modelling-challenges and solutions.* New York: Springer.

Sofer, M., & Schnell, I. (2002). Over-and under-embeddedness; failures in developing mixed embeddedness among Israeli Arab entrepreneurs. In *Embedded enterprise and social capital: International perspectives* (pp. 207–224). Aldershot: Ashgate Publishing Limited.

Ter Wal, A. L., & Boschma, R. A. (2009). Applying social network analysis in economic geography: Framing some key analytic issues. *The Annals of Regional Science, 43*(3), 739–756.

Thiele, J. C., Kurth, W., & Grimm, V. (2014). Facilitating parameter estimation and sensitivity analysis of agent-based models: A cookbook using NetLogo and R. *Journal of Artificial Societies and Social Simulation, 17*(3), 11.

Ting Helena Chiu, Y. (2008). How network competence and network location influence innovation performance. *Journal of Business & Industrial Marketing, 24*(1), 46–55.

Uzzi, B. (1997). Social structure and competition in interfirm networks: The paradox of embeddedness. *Administrative Science Quarterly, 42*, 35–67.

Vasudeva, G., Zaheer, A., & Hernandez, E. (2013). The embeddedness of networks: Institutions, structural holes, and innovativeness in the fuel cell industry. *Organization Science, 24*(3), 645–663.

Veugelers, R., & Cassiman, B. (2005). R&D cooperation between firms and universities. Some empirical evidence from Belgian manufacturing. *International Journal of Industrial Organization, 23*(5), 355–379.

Wajsman, N., Thumm, N., Kazimierczak, M., Lazaridis, G., Arias Burgos, C., Domanico, F., García Valero, F., Boedt, G., Garanasvili, A., & Mihailescu, A. (2013). Intellectual property rights intensive industries: Contribution to economic performance and employment in the European Union. In *Industry-level analysis report.* Munich /Alicante: European Patent Office and Office for Harmonization in the Internal Market.

Wanzenböck, I., & Piribauer, P. (2016). R&D networks and regional knowledge production in Europe: Evidence from a space-time model. *Papers in Regional Science., 97*, S1–S24.

Wasserman, S., & Faust, K. (1994). *Social network analysis: Methods and applications.* Cambridge: Cambridge University Press.

Whittington, K. B., Owen-Smith, J., & Powell, W. W. (2009). Networks, propinquity, and innovation in knowledge-intensive industries. *Administrative Science Quarterly, 54*(1), 90–122.

Zaheer, A., & Bell, G. G. (2005). Benefiting from network position: Firm capabilities, structural holes, and performance. *Strategic Management Journal, 26*(9), 809–825.

Index

© Springer Nature Switzerland AG 2019
D. Payne et al. (eds.), *Social Simulation for a Digital Society*, Springer
Proceedings in Complexity, https://doi.org/10.1007/978-3-030-30298-6

Printed in the United States
By Bookmasters